Bioenergy: Principles, Technology and Applications

Bioenergy: Principles, Technology and Applications

Edited by
Sterling Macmillan

▤ Larsen & Keller
www.larsen-keller.com

Bioenergy: Principles, Technology and Applications
Edited by Sterling Macmillan
ISBN: 978-1-63549-043-5 (Hardback)

© 2017 Larsen & Keller

⊟ Larsen & Keller

Published by Larsen and Keller Education,
5 Penn Plaza,
19th Floor,
New York, NY 10001, USA

Cataloging-in-Publication Data

Bioenergy : principles, technology and applications / edited by Sterling Macmillan.
   p. cm.
Includes bibliographical references and index.
ISBN 978-1-63549-043-5
1. Biomass energy. 2. Renewable energy sources. I. Macmillan, Sterling.
TP339 .B54 2017
662.88--dc23

The publisher's policy is to use permanent paper from mills that operate a sustainable forestry policy. Furthermore, the publisher ensures that the text paper and cover boards used have met acceptable environmental accreditation standards.

Printed and bound in the United States of America.

For more information regarding Larsen and Keller Education and its products, please visit the publisher's website www.larsen-keller.com

Table of Contents

Preface   **VII**

Chapter 1 **Introduction to Bioenergy**   **1**

Chapter 2 **Renewable Energy: An Overview**   **6**
    a. Renewable Energy   6

Chapter 3 **Biomass: An Integrated Study**   **105**
    a. Biomass   105
    b. Renewable Natural Gas   114
    c. Biochar   117

Chapter 4 **Fuels: Sources of Bioenergy**   **124**
    a. Wood   124
    b. Straw   142
    c. Manure   147
    d. Sugarcane   151

Chapter 5 **Different Types of Efficient Fuels**   **167**
    a. Biogas   167
    b. Biofuel   179
    c. Pellet Fuel   195

Chapter 6 **Biodiesel and its Environmental Impacts**   **204**
    a. Biodiesel   204
    b. Environmental Impact of Biodiesel   229

Chapter 7 **Bioenergy with Carbon Capture and Storage**   **233**

Chapter 8 **Various Aspects of Bioenergy**   **239**
    a. Sustainable Development   239
    b. Sustainability   256

Permissions

Index

Preface

This book provides comprehensive insights into the field of bioenergy. It attempts to understand the multiple branches that fall under this discipline and how such concepts have practical applications. Bioenergy refers to a form of renewable energy which is produced by using biological resources. The resources used are biomass, wood, straw, sugar cane, manure and other agricultural products. This book also describes in detail the positive and the negative impacts of bioenergy. Most of the topics introduced in this textbook cover new techniques and the applications of bioenergy. Different approaches, evaluations and methodologies on this subject have been included in it. For someone with an interest and eye for detail, this book covers the most significant topics in the field of bioenergy. This text will serve as a reference to a broad spectrum of readers.

A detailed account of the significant topics covered in this book is provided below:

Chapter 1- The energy produced from biological sources is known as bioenergy. As a fuel, bioenergy includes wood, straw, manure, sugarcane and some other products. The chapter on bioenergy an insightful focus, keeping in mind the subject matter.

Chapter 2- The energy that is collected from resources is known as renewable energy. Sunlight, wind, rain and geothermal heat are perfect examples of renewable energy. The trends and technologies used in renewable energy are wind power, tidal power, hydropower, solar energy and geothermal energy. This section is an overview of the subject matter incorporating all the major aspects of renewable energy.

Chapter 3- Biomass is the organic matter that is derived from living organisms. It usually refers to plants or plant-based materials. The following text focuses on topics such as renewable natural gas and biochar. The topics discussed in the chapter are of great importance to broaden the existing knowledge on biomass.

Chapter 4- This chapter focuses on the sources of bioenergy. The fuels explained in this section are wood, straw, manure and sugarcane. Wood has been used as a fuel for thousands of years whereas straw is an agricultural by-product and makes up for about half of the yield of cereal crops. The topics discussed in this chapter are of vital importance and provides a better understanding of bioenergy.

Chapter 5- Efficient fuels are best understood in confluence with the major types listed in the following section. The different types of efficient fuels discussed in this text are biogas, biofuel and pellet fuel. Biogas is the combination of different gases, produced by the breakdown of organic matter in the absence of oxygen. This chapter has been carefully written to provide an easy understanding of the different types of efficient fuels.

Chapter 6- Biodiesel usually refers to a vegetable oil; it is typically made by chemically reacting lipids. Biodiesel is used with the intention of reducing greenhouse gas. Whether it certainly does cause less environmental issues or not depends on many factors. The diverse applications of biodiesel in the current scenario have been thoroughly discussed in this section.

Chapter 7- Bio-energy with carbon capture and storage produces negative carbon dioxide emissions by combining the use of bioenergy with geologic carbon capture and storage. This chapter will provide the readers with an integrated understanding of carbon capture and storage.

Chapter 8- The aspects of bioenergy are sustainable development and sustainability. Sustainable development is the procedure of meeting human development goals; this is achieved by sustaining natural resources and protecting the environment. The section strategically encompasses and incorporates the major aspects of bioenergy, providing the reader with a complete understanding on the respective subject.

It gives me an immense pleasure to thank our entire team for their efforts. Finally in the end, I would like to thank my family and colleagues who have been a great source of inspiration and support.

Editor

Introduction to Bioenergy

The energy produced from biological sources is known as bioenergy. As a fuel, bioenergy includes wood, straw, manure, sugarcane and some other products. The chapter on bioenergy an insightful focus, keeping in mind the subject matter.

Bioenergy is renewable energy made available from materials derived from biological sources. Biomass is any organic material which has stored sunlight in the form of chemical energy. As a fuel it may include wood, wood waste, straw, manure, sugarcane, and many other by products from a variety of agricultural processes. By 2010, there was 35 GW (47,000,000 hp) of globally installed bioenergy capacity for electricity generation, of which 7 GW (9,400,000 hp) was in the United States.

A Stirling engine capable of producing electricity from biomass combustion heat

In its most narrow sense it is a synonym to biofuel, which is fuel derived from biological sources. In its broader sense it includes biomass, the biological material used as a biofuel, as well as the social, economic, scientific and technical fields associated with using biological sources for energy. This is a common misconception, as bioenergy is the energy extracted from the biomass, as the biomass is the fuel and the bioenergy is the energy contained in the fuel

There is a slight tendency for the word *bioenergy* to be favoured in Europe compared with *biofuel* in America.

Solid Biomass

One of the advantages of biomass fuel is that it is often a by-product, residue or waste-product of other processes, such as farming, animal husbandry and forestry. In theory this means there is no competition between fuel and food production, although this is not always the case. Land use, existing biomass industries and relevant conversion technologies must be considered when evaluating suitability of developing biomass as feedstock for energy.

Simple use of biomass fuel (Combustion of wood for heat).

Biomass is the material derived from recently living organisms, which includes plants, animals and their byproducts. Manure, garden waste and crop residues are all sources of biomass. It is a renewable energy source based on the carbon cycle, unlike other natural resources such as petroleum, coal, and nuclear fuels. Another source includes Animal waste, which is a persistent and unavoidable pollutant produced primarily by the animals housed in industrial-sized farms.

There are also agricultural products specifically being grown for biofuel production. These include corn, and soybeans and to some extent willow and switchgrass on a pre-commercial research level, primarily in the United States; rapeseed, wheat, sugar beet, and willow (15,000 ha or 37,000 acres in Sweden) primarily in Europe; sugarcane in Brazil; palm oil and miscanthus in Southeast Asia; sorghum and cassava in China; and jatropha in India. Hemp has also been proven to work as a biofuel. Biodegradable outputs from industry, agriculture, forestry and households can be used for biofuel production, using e.g. anaerobic digestion to produce biogas, gasification to produce syngas or by direct combustion. Examples of biodegradable wastes include straw, timber, manure, rice husks, sewage, and food waste. The use of biomass fuels can therefore contribute to waste management as well as fuel security and help to prevent or slow down climate change, although alone they are not a comprehensive solution to these problems.

Biomass can be converted to other usable forms of energy like methane gas or transportation fuels like ethanol and biodiesel. Rotting garbage, and agricultural and human waste, all release methane gas—also called "landfill gas" or "biogas." Crops, such as corn and sugar cane, can be fermented to produce the transportation fuel, ethanol. Bio-diesel, another transportation fuel, can be produced from left-over food products like vegetable oils and animal fats. Also, Biomass to liquids (BTLs) and cellulosic ethanol are still under research.

Sewage Biomass

A new bioenergy sewage treatment process aimed at developing countries is now on the horizon; the Omni Processor is a self-sustaining process which uses the sewerage solids as fuel to convert sewage waste water into drinking water and electrical energy.

Electricity Generation from Biomass

The biomass used for electricity production ranges by region. Forest byproducts, such as wood residues, are popular in the United States. Agricultural waste is common in Mauritius (sugar cane residue) and Southeast Asia (rice husks). Animal husbandry res-idues, such as poultry litter, is popular in the UK.

Electricity from Sugarcane Bagasse in Brazil

Sucrose accounts for little more than 30% of the chemical energy stored in the mature plant; 35% is in the leaves and stem tips, which are left in the fields during harvest, and 35% are in the fibrous material (bagasse) left over from pressing.

Sugarcane (*Saccharum officinarum*) plantation ready for harvest, Ituverava, São Paulo State. Brazil.

The production process of sugar and ethanol in Brazil takes full advantage of the energy stored in sugarcane. Part of the bagasse is currently burned at the mill to provide heat for distillation and electricity to run the machinery. This allows ethanol plants to be en-ergetically self-sufficient and even sell surplus electricity to utilities; current production is 600 MW (800,000 hp) for self-use and 100 MW (130,000 hp) for sale. This second-ary activity is expected to boom now that utilities have been induced to pay "fair price "(about US$10/GJ or US$0.036/kWh) for 10 year contracts. This is approximately half of what the World Bank considers the reference price for investing in similar projects.

The energy is especially valuable to utilities because it is produced mainly in the dry season when hydroelectric dams are running low. Estimates of potential power generation from bagasse range from 1,000 to 9,000 MW (1,300,000 to 12,100,000 hp), depending on technology. Higher estimates assume gasification of biomass, replacement of current low-pressure steam boilers and turbines by high-pressure ones, and use of harvest trash currently left behind in the fields. For comparison, Brazil's Angra I nuclear plant generates 657 MW (881,000 hp).

A sugar/ethanol plant located in Piracicaba, São Paulo State. This plant produces the electricity it needs from bagasse residuals from sugarcane left over by the milling process, and it sells the surplus electricity to the public grid.

Presently, it is economically viable to extract about 288 MJ of electricity from the residues of one tonne of sugarcane, of which about 180 MJ are used in the plant itself. Thus a medium-size distillery processing 1,000,000 tonnes (980,000 long tons; 1,100,000 short tons) of sugarcane per year could sell about 5 MW (6,700 hp) of surplus electricity. At current prices, it would earn US$18 million from sugar and ethanol sales, and about US$1 million from surplus electricity sales. With advanced boiler and turbine technology, the electricity yield could be increased to 648 MJ per tonne of sugarcane, but current electricity prices do not justify the necessary investment. (According to one report, the World Bank would only finance investments in bagasse power generation if the price were at least US$19/GJ or US$0.068/kWh.)

Bagasse burning is environmentally friendly compared to other fuels like oil and coal. Its ash content is only 2.5% (against 30–50% of coal), and it contains very little sulfur. Since it burns at relatively low temperatures, it produces little nitrous oxides. Moreover, bagasse is being sold for use as a fuel (replacing heavy fuel oil) in various industries, including citrus juice concentrate, vegetable oil, ceramics, and tyre recycling. The state of São Paulo alone used 2,000,000 tonnes (1,970,000 long tons; 2,200,000 short tons), saving about US$35 million in fuel oil imports.

Researchers working with cellulosic ethanol are trying to make the extraction of ethanol from sugarcane bagasse and other plants viable on an industrial scale.

Electricity from Electrogenic Micro-organisms

Another form of bioenergy can be attained from microbial fuel cells, in which chemical energy stored in wastewater or soil is converted directly into electrical energy via the metabolic processes of electrogenic micro-organisms. The power generation capability of this technology has not been economical to date, however, and this technology has found more utility for chemical treatment processes and student education.

Environmental Impact

Some forms of forest bioenergy have recently come under fire from a number of environmental organizations, including Greenpeace and the Natural Resources Defense Council, for the harmful impacts they can have on forests and the climate. Greenpeace recently released a report entitled Fuelling a BioMess which outlines their concerns around forest bioenergy. Because any part of the tree can be burned, the harvesting of trees for energy production encourages Whole-Tree Harvesting, which removes more nutrients and soil cover than regular harvesting, and can be harmful to the long-term health of the forest. In some jurisdictions, forest biomass is increasingly consisting of elements essential to functioning forest ecosystems, including standing trees, naturally disturbed forests and remains of traditional logging operations that were previously left in the forest. Environmental groups also cite recent scientific research which has found that it can take many decades for the carbon released by burning biomass to be recaptured by regrowing trees, and even longer in low productivity areas; furthermore, logging operations may disturb forest soils and cause them to release stored carbon. In light of the pressing need to reduce greenhouse gas emissions in the short term in order to mitigate the effects of climate change, a number of environmental groups are opposing the large-scale use of forest biomass in energy production.

References

- "NRDC fact sheet lays out biomass basics, campaign calls for action to tell EPA our forests aren't fuel". nrdc.org. Retrieved 28 February 2015.

- Frauke Urban and Tom Mitchell 2011. Climate change, disasters and electricity generation. London: Overseas Development Institute and Institute of Development Studies.

Renewable Energy: An Overview

The energy that is collected from resources is known as renewable energy. Sunlight, wind, rain and geothermal heat are perfect examples of renewable energy. The trends and technologies used in renewable energy are wind power, tidal power, hydropower, solar energy and geothermal energy. This section is an overview of the subject matter incorporating all the major aspects of renewable energy.

Renewable Energy

Renewable energy is generally defined as energy that is collected from resources which are naturally replenished on a human timescale, such as sunlight, wind, rain, tides, waves, and geothermal heat. Renewable energy often provides energy in four important areas: electricity generation, air and water heating/cooling, transportation, and rural (off-grid) energy services.

Wind, solar, and hydroelectricity are three emerging renewable sources of energy

Based on REN21's 2016 report, renewables contributed 19.2% to humans' global energy consumption and 23.7% to their generation of electricity in 2014 and 2015, respectively. This energy consumption is divided as 8.9% coming from traditional biomass, 4.2% as heat energy (modern biomass, geothermal and solar heat), 3.9% hydro electricity and 2.2% is electricity from wind, solar, geothermal, and biomass. Worldwide investments in renewable technologies amounted to more than US$286 billion in 2015, with countries like China and the United States heavily investing in wind, hydro, solar and biofuels. Globally, there are an estimated 7.7 million jobs associated with the renewable energy industries, with solar photovoltaics being the largest renewable employer.

Global public support for energy sources

"Please indicate whether you strongly support, somewhat support, somewhat oppose, or strongly oppose each way of producing energy"

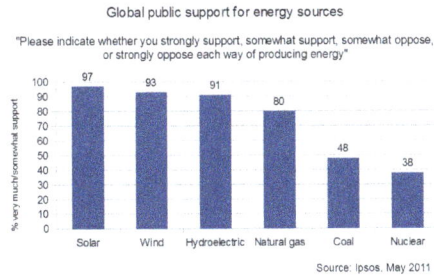

Source: Ipsos, May 2011

Global public support for different energy sources (2011) based on a poll by Ipsos Global @dvisor

Renewable energy resources exist over wide geographical areas, in contrast to other energy sources, which are concentrated in a limited number of countries. Rapid deployment of renewable energy and energy efficiency is resulting in significant energy security, climate change mitigation, and economic benefits. The results of a recent review of the literature concluded that as greenhouse gas (GHG) emitters begin to be held liable for damages resulting from GHG emissions resulting in climate change, a high value for liability mitigation would provide powerful incentives for deployment of renewable energy technologies. In international public opinion surveys there is strong support for promoting renewable sources such as solar power and wind power. At the national level, at least 30 nations around the world already have renewable energy contributing more than 20 percent of energy supply. National renewable energy markets are projected to continue to grow strongly in the coming decade and beyond. Some places and at least two countries, Iceland and Norway generate all their electricity using renewable energy already, and many other countries have the set a goal to reach 100% renewable energy in the future. For example, in Denmark the government decided to switch the total energy supply (electricity, mobility and heating/cooling) to 100% renewable energy by 2050.

While many renewable energy projects are large-scale, renewable technologies are also suited to rural and remote areas and developing countries, where energy is often crucial in human development. United Nations' Secretary-General Ban Ki-moon has said that renewable energy has the ability to lift the poorest nations to new levels of prosperity. As most of renewables provide electricity, renewable energy deployment is often applied in conjunction with further electrification, which has several benefits: For example, electricity can be converted to heat without losses and even reach higher temperatures than fossil fuels, can be converted into mechanical energy with high efficiency and is clean at the point of consumpion. In addition to that electrification with renewable energy is much more efficient and therefore leads to a significant reduction in primary energy requirements, because most renewables don't have a steam cycle with high losses (fossil power plants usually have losses of 40 to 65%).

Overview

Renewable energy flows involve natural phenomena such as sunlight, wind, tides, plant growth, and geothermal heat, as the International Energy Agency explains:

Renewable energy is derived from natural processes that are replenished constantly. In its various forms, it derives directly from the sun, or from heat generated deep within the earth. Included in the definition is electricity and heat generated from solar, wind, ocean, hydropower, biomass, geothermal resources, and biofuels and hydrogen derived from renewable resources.

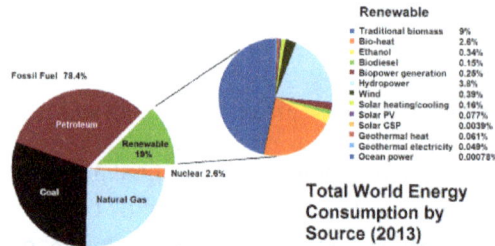

World energy consumption by source. Renewables accounted for 19% in 2012.

PlanetSolar, the world's largest solar-powered boat and the first ever solar electric vehicle to circumnavigate the globe (in 2012)

Renewable energy resources and significant opportunities for energy efficiency exist over wide geographical areas, in contrast to other energy sources, which are concentrated in a limited number of countries. Rapid deployment of renewable energy and energy efficiency, and technological diversification of energy sources, would result in significant energy security and economic benefits. It would also reduce environmental pollution such as air pollution caused by burning of fossil fuels and improve public health, reduce premature mortalities due to pollution and save associated health costs that amount to several hundred billion dollars annually only in the United States. Renewable energy sources, that derive their energy from the sun, either directly or indirectly, such as hydro and wind, are expected to be capable of supplying humanity energy for almost another 1 billion years, at which point the predicted increase in heat from the sun is expected to make the surface of the earth too hot for liquid water to exist.

Climate change and global warming concerns, coupled with high oil prices, peak oil, and increasing government support, are driving increasing renewable energy legislation, incentives and commercialization. New government spending, regulation and policies helped the industry weather the global financial crisis better than many other sectors. According to a 2011 projection by the International Energy Agency, solar power generators may produce most of the world's electricity within 50 years, reducing the emissions of greenhouse gases that harm the environment.

As of 2011, small solar PV systems provide electricity to a few million households, and micro-hydro configured into mini-grids serves many more. Over 44 million households use biogas made in household-scale digesters for lighting and/or cooking, and more than 166 million households rely on a new generation of more-efficient biomass cook-stoves. United Nations' Secretary-General Ban Ki-moon has said that renewable energy has the ability to lift the poorest nations to new levels of prosperity. At the national level, at least 30 nations around the world already have renewable energy contributing more than 20% of energy supply. National renewable energy markets are projected to continue to grow strongly in the coming decade and beyond, and some 120 countries have various policy targets for longer-term shares of renewable energy, including a 20% target of all electricity generated for the European Union by 2020. Some countries have much higher long-term policy targets of up to 100% renewables. Outside Europe, a diverse group of 20 or more other countries target renewable energy shares in the 2020–2030 time frame that range from 10% to 50%.

Renewable energy often displaces conventional fuels in four areas: electricity generation, hot water/space heating, transportation, and rural (off-grid) energy services:

Power Generation

Renewable hydroelectric energy provides 16.3% of the worlds electricity. When hydroelectric is combined with other renewables such as wind, geothermal, solar, biomass and waste: together they make the "renewables" total, 21.7% of electricity generation worldwide as of 2013. Renewable power generators are spread across many countries, and wind power alone already provides a significant share of electricity in some areas: for example, 14% in the U.S. state of Iowa, 40% in the northern German state of Schleswig-Holstein, and 49% in Denmark. Some countries get most of their power from renewables, including Iceland (100%), Norway (98%), Brazil (86%), Austria (62%), New Zealand (65%), and Sweden (54%).

Heating

Solar water heating makes an important contribution to renewable heat in many countries, most notably in China, which now has 70% of the global total (180 GWth). Most of these systems are installed on multi-family apartment buildings and meet a portion of the hot water needs of an estimated 50–60 million households in China. Worldwide, total installed solar water heating systems meet a portion of the water heating needs of over 70 million households. The use of biomass for heating continues to grow as well. In Sweden, national use of biomass energy has surpassed that of oil. Direct geothermal for heating is also growing rapidly. The newest addition to Heating is from Geothermal Heat

Pumps which provide both heating and cooling, and also flatten the electric demand curve and are thus an increasing national priority.

Transportation

A bus fueled by biodiesel

Bioethanol is an alcohol made by fermentation, mostly from carbohydrates produced in sugar or starch crops such as corn, sugarcane, or sweet sorghum. Cellulosic biomass, derived from non-food sources such as trees and grasses is also being developed as a feedstock for ethanol production. Ethanol can be used as a fuel for vehicles in its pure form, but it is usually used as a gasoline additive to increase octane and improve vehicle emissions. Bioethanol is widely used in the USA and in Brazil. Biodiesel can be used as a fuel for vehicles in its pure form, but it is usually used as a diesel additive to reduce levels of particulates, carbon monoxide, and hydrocarbons from diesel-powered vehicles. Biodiesel is produced from oils or fats using transesterification and is the most common biofuel in Europe.

A solar vehicle is an electric vehicle powered completely or significantly by direct solar energy. Usually, photovoltaic (PV) cells contained in solar panels convert the sun's energy directly into electric energy. The term "solar vehicle" usually implies that solar energy is used to power all or part of a vehicle's propulsion. Solar power may be also used to provide power for communications or controls or other auxiliary functions. Solar powered boats have mainly been limited to rivers and canals, but in 2007 an experimental 14m catamaran, the Sun21 sailed the Atlantic from Seville to Miami, and from there to New York. It was the first crossing of the Atlantic powered only by solar. Solar vehicles are not sold as practical day-to-day transportation devices at present, but are primarily demonstration vehicles and engineering exercises, often sponsored by government agencies. However, indirectly solar-charged vehicles are widespread and solar boats are available commercially.

History

Prior to the development of coal in the mid 19th century, nearly all energy used was renewable. Almost without a doubt the oldest known use of renewable energy, in the form of traditional biomass to fuel fires, dates from 790,000 years ago. Use of biomass for fire did not become commonplace until many hundreds of thousands of years later, sometime between 200,000 and 400,000 years ago. Probably the second oldest usage of renewable energy is harnessing the wind in order to drive ships over water. This practice can be traced back some 7000 years, to ships on the Nile. Moving into the time of recorded history, the primary sources of traditional renewable energy were human labor, animal power, water power, wind, in grain crushing windmills, and firewood, a traditional biomass. A graph of energy use in the United States up until 1900 shows oil and natural gas with about the same importance in 1900 as wind and solar played in 2010.

In the 1860s and '70s there were already fears that civilization would run out of fossil fuels and the need was felt for a better source. In 1873 Professor Augustine Mouchot wrote:

The time will arrive when the industry of Europe will cease to find those natural resources, so necessary for it. Petroleum springs and coal mines are not inexhaustible but are rapidly diminishing in many places. Will man, then, return to the power of water and wind? Or will he emigrate where the most powerful source of heat sends its rays to all? History will show what will come.

In 1885, Werner von Siemens, commenting on the discovery of the photovoltaic effect in the solid state, wrote:

In conclusion, I would say that however great the scientific importance of this discovery may be, its practical value will be no less obvious when we reflect that the supply of solar energy is both without limit and without cost, and that it will continue to pour down upon us for countless ages after all the coal deposits of the earth have been exhausted and forgotten.

Max Weber mentioned the end of fossil fuel in the concluding paragraphs of his Die protestantische Ethik und der Geist des Kapitalismus, published in 1905.

Development of solar engines continued until the outbreak of World War I. The importance of solar energy was recognized in a 1911 Scientific American article: "in the far distant future, natural fuels having been exhausted [solar power] will remain as the only means of existence of the human race".

The theory of peak oil was published in 1956. In the 1970s environmentalists promoted the development of renewable energy both as a replacement for the eventual depletion of oil, as well as for an escape from dependence on oil, and the first electricity generating wind turbines appeared. Solar had long been used for heating and cooling, but solar panels were too costly to build solar farms until 1980.

The IEA 2014 World Energy Outlook projects a growth of renewable energy supply from 1,700 gigawatts in 2014 to 4,550 gigawatts in 2040. Fossil fuels received about $550 billion in subsidies in 2013, compared to $120 billion for all renewable energies.

Mainstream Technologies

Wind Power

Airflows can be used to run wind turbines. Modern utility-scale wind turbines range from around 600 kW to 5 MW of rated power, although turbines with rated output of 1.5–3 MW have become the most common for commercial use; the power available from the wind is a function of the cube of the wind speed, so as wind speed increases, power output increases up to the maximum output for the particular turbine. Areas where winds are stronger and more constant, such as offshore and high altitude sites, are preferred locations for wind farms. Typically full load hours of wind turbines vary between 16 and 57 percent annually, but might be higher in particularly favorable offshore sites.

The 845 MW Shepherds Flat Wind Farm near Arlington, Oregon, USA

Globally, the long-term technical potential of wind energy is believed to be five times total current global energy production, or 40 times current electricity demand, assuming all practical barriers needed were overcome. This would require wind turbines to be installed over large areas, particularly in areas of higher wind resources, such as offshore. As offshore wind speeds average ~90% greater than that of land, so offshore resources can contribute substantially more energy than land stationed turbines. In 2014 global wind generation was 706 terawatt-hours or 3% of the worlds total electricity.

Hydropower

The Three Gorges Dam on the Yangtze River in China

In 2015 hydropower generated 16.6% of the worlds total electricity and 70% of all renewable electricity. Since water is about 800 times denser than air, even a slow flowing stream of water, or moderate sea swell, can yield considerable amounts of energy. There are many forms of water energy:

- Historically hydroelectric power came from constructing large hydroelectric dams and reservoirs, which are still popular in third world countries. The largest of which is the Three Gorges Dam(2003) in China and the Itaipu Dam(1984) built by Brazil and Paraguay.

- Small hydro systems are hydroelectric power installations that typically produce up to 50 MW of power. They are often used on small rivers or as a low impact development on larger rivers. China is the largest producer of hydroelectricity in the world and has more than 45,000 small hydro installations.

- Run-of-the-river hydroelectricity plants derive kinetic energy from rivers without the creation of a large reservoir. This style of generation may still produce a large amount of electricity, such as the Chief Joseph Dam on the Columbia river in the United States.

Hydropower is produced in 150 countries, with the Asia-Pacific region generating 32 percent of global hydropower in 2010. For counties having the largest percentage of electricity from renewables, the top 50 are primarily hydroelectric. China is the largest hydroelectricity producer, with 721 terawatt-hours of production in 2010, representing around 17 percent of domestic electricity use. There are now three hydroelectricity stations larger than 10 GW: the Three Gorges Dam in China, Itaipu Dam across the Brazil/Paraguay border, and Guri Dam in Venezuela.

Wave power, which captures the energy of ocean surface waves, and tidal power, converting the energy of tides, are two forms of hydropower with future potential; however, they are not yet widely employed commercially. A demonstration project operated by the Ocean Renewable Power Company on the coast of Maine, and connected to the grid, harnesses tidal power from the Bay of Fundy, location of world's highest tidal flow. Ocean thermal energy conversion, which uses the temperature difference between cooler deep and warmer surface waters, has currently no economic feasibility.

Solar Energy

Solar energy, radiant light and heat from the sun, is harnessed using a range of ever-evolving technologies such as solar heating, photovoltaics, concentrated solar power (CSP), concentrator photovoltaics (CPV), solar architecture and artificial photosynthesis. Solar technologies are broadly characterized as either passive solar or active solar depending on the way they capture, convert and distribute solar energy. Passive solar techniques include orienting a building to the Sun, selecting materials with favorable thermal mass or light dispersing properties, and designing spaces that naturally circulate air. Active solar technologies encompass solar thermal energy, using solar collectors for heating, and solar power, converting sunlight into electricity either directly using photovoltaics (PV), or indirectly using concentrated solar power (CSP).

Satellite image of the 550-megawatt Topaz Solar Farm in California, USA

A photovoltaic system converts light into electrical direct current (DC) by taking advantage of the photoelectric effect. Solar PV has turned into a multi-billion, fast-growing industry, continues to improve its cost-effectiveness, and has the most potential of any renewable technologies together with CSP. Concentrated solar power (CSP) systems use lenses or mirrors and tracking systems to focus a large area of sunlight into a small beam. Commercial concentrated solar power plants were first developed in the 1980s. CSP-Stirling has by far the highest efficiency among all solar energy technologies.

In 2011, the International Energy Agency said that "the development of affordable, inexhaustible and clean solar energy technologies will have huge longer-term benefits. It will increase countries' energy security through reliance on an indigenous, inexhaustible and mostly import-independent resource, enhance sustainability, reduce pollution, lower the costs of mitigating climate change, and keep fossil fuel prices lower than otherwise. These advantages are global. Hence the additional costs of the incentives for early deployment should be considered learning investments; they must be wisely spent and need to be widely shared". In 2014 global solar generation was 186 terawatt-hours, slightly less than 1% of the worlds total grid electricity.

Geothermal Energy

Steam rising from the Nesjavellir Geothermal Power Station in Iceland

High Temperature Geothermal energy is from thermal energy generated and stored in the Earth. Thermal energy is the energy that determines the temperature of matter. Earth's geothermal energy originates from the original formation of the planet and from radioactive decay of minerals (in currently uncertain but possibly roughly equal proportions). The geothermal gradient, which is the difference in temperature between the core of the planet and its surface, drives a continuous conduction of thermal energy in the form of heat from the core to the surface. The adjective *geothermal* originates from the Greek roots *geo*, meaning earth, and *thermos*, meaning heat.

The heat that is used for geothermal energy can be from deep within the Earth, all the way down to Earth's core – 4,000 miles (6,400 km) down. At the core, temperatures may reach over 9,000 °F (5,000 °C). Heat conducts from the core to surrounding rock. Extremely high temperature and pressure cause some rock to melt, which is commonly known as magma. Magma convects upward since it is lighter than the solid rock. This magma then heats rock and water in the crust, sometimes up to 700 °F (371 °C).

From hot springs, geothermal energy has been used for bathing since Paleolithic times and for space heating since ancient Roman times, but it is now better known for electricity generation.

Low Temperature Geothermal refers to the use of the outer crust of the earth as a Thermal Battery to facilitate Renewable thermal energy for heating and cooling buildings, and other refrigeration and industrial uses. In this form of Geothermal, a Geothermal Heat Pump and Ground-coupled heat exchanger are used together to move heat energy into the earth (for cooling) and out of the earth (for heating) on a varying seasonal basis. Low temperature Geothermal (generally referred to as "GHP") is an increasingly important renewable technology because it both reduces total annual energy loads associated with heating and cooling, and it also flattens the electric demand curve eliminating the extreme summer and winter peak electric supply requirements. Thus Low Temperature Geothermal/GHP is becoming an increasing national priority with multiple tax credit support and focus as part of the ongoing movement toward Net Zero Energy. New York City has even just passed a law to require GHP anytime is shown to be economical with 20 year financing including the Socialized Cost of Carbon.

Bio Energy

Sugarcane plantation to produce ethanol in Brazil

Biomass is biological material derived from living, or recently living organisms. It most often refers to plants or plant-derived materials which are specifically called lignocellulosic biomass. As an energy source, biomass can either be used directly via combustion to produce heat, or indirectly after converting it to various forms of biofuel. Conversion of biomass to biofuel can be achieved by different methods which are broadly classified into: *thermal*, *chemical*, and *biochemical* methods. Wood remains the largest biomass energy source today; examples include forest residues – such as dead trees, branches and tree stumps –, yard clippings, wood chips and even municipal solid waste. In the second sense, biomass includes plant or animal matter that can be converted into fibers or other industrial chemicals, including biofuels. Industrial biomass can be grown from numerous types of plants, including miscanthus, switchgrass, hemp, corn, poplar, willow, sorghum, sugarcane, bamboo, and a variety of tree species, ranging from eucalyptus to oil palm (palm oil).

A CHP power station using wood to supply 30,000 households in France

Plant energy is produced by crops specifically grown for use as fuel that offer high biomass output per hectare with low input energy. Some examples of these plants are wheat, which typically yield 7.5–8 tonnes of grain per hectare, and straw, which typically yield 3.5–5 tonnes per hectare in the UK. The grain can be used for liquid transportation fuels while the straw can be burned to produce heat or electricity. Plant biomass can also be degraded from cellulose to glucose through a series of chemical treatments, and the resulting sugar can then be used as a first generation biofuel.

Biomass can be converted to other usable forms of energy like methane gas or transportation fuels like ethanol and biodiesel. Rotting garbage, and agricultural and human waste, all release methane gas – also called landfill gas or biogas. Crops, such as corn and sugarcane, can be fermented to produce the transportation fuel, ethanol. Biodiesel, another transportation fuel, can be produced from left-over food products like vegetable oils and animal fats. Also, biomass to liquids (BTLs) and cellulosic ethanol are still under research. There is a great deal of research involving algal fuel or algae-derived biomass due to the fact that it's a non-food resource and can be produced at rates 5 to 10 times those of other types of land-based agriculture, such as corn and soy. Once harvested, it can be fermented to produce biofuels such as ethanol, butanol, and methane, as well as biodiesel and hydrogen. The biomass used for electricity generation varies by region. Forest by-products, such as wood residues, are common in the United States. Agricultural waste is common in Mauritius (sugar cane residue) and Southeast Asia (rice husks). Animal husbandry residues, such as poultry litter, are common in the United Kingdom.

Biofuels include a wide range of fuels which are derived from biomass. The term covers solid, liquid, and gaseous fuels. Liquid biofuels include bioalcohols, such as bioethanol, and oils, such as biodiesel. Gaseous biofuels include biogas, landfill gas and synthetic gas. Bioethanol is an alcohol made by fermenting the sugar components of plant materials and it is made mostly from sugar and starch crops. These include maize, sugarcane and, more recently, sweet sorghum. The latter crop is particularly suitable for growing in dryland conditions, and is being investigated by International Crops Research Institute for the Semi-Arid Tropics for its potential to provide fuel, along with food and animal feed, in arid parts of Asia and Africa.

With advanced technology being developed, cellulosic biomass, such as trees and grasses, are also used as feedstocks for ethanol production. Ethanol can be used as a fuel for vehicles in its pure form, but it is usually used as a gasoline additive to increase octane and improve vehicle emissions. Bioethanol is widely used in the United States and in Brazil. The energy costs for producing bio-ethanol are almost equal to, the energy yields from bio-ethanol. However, according to the European Environment Agency, biofuels do not address global warming concerns. Biodiesel is made from vegetable oils, animal fats or recycled greases. It can be used as a fuel for vehicles in its pure form, or more commonly as a diesel additive to reduce levels of particulates, carbon monoxide, and hydrocarbons from diesel-powered vehicles. Biodiesel is produced from oils or fats using transesterification and is the most common biofuel in Europe. Biofuels provided 2.7% of the world's transport fuel in 2010.

Biomass, biogas and biofuels are burned to produce heat/power and in doing so harm the environment. Pollutants such as sulphurous oxides (SO_x), nitrous oxides (NO_x), and particulate matter (PM) are produced from the combustion of biomass; the World Health Organisation estimates that 7 million premature deaths are caused each year by air pollution. Biomass combustion is a major contributor. The life cycle of the plants is sustainable, the lives of people less so.

Energy Storage

Energy storage is a collection of methods used to store electrical energy on an electrical power grid, or off it. Electrical energy is stored during times when production (especially from intermittent power plants such as renewable electricity sources such as wind power, tidal power, solar power) exceeds consumption, and returned to the grid when production falls below consumption. Water pumped into a hydroelectric dam is the largest form of power storage.

Market and Industry Trends

Growth of Renewables

From the end of 2004, worldwide renewable energy capacity grew at rates of 10–60% annually for many technologies. For wind power and many other renewable technologies, growth accelerated in 2009 relative to the previous four years. More wind power capacity was added during 2009 than any other renewable technology. However, grid-connected PV increased the fastest of all renewables technologies, with a 60% annual average growth rate. In 2010, renewable power constituted about a third of the newly built power generation capacities.

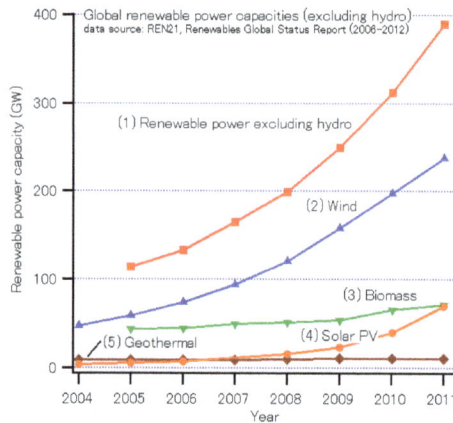

Global growth of renewables through 2011

Projections vary, but scientists have advanced a plan to power 100% of the world's energy with wind, hydroelectric, and solar power by the year 2030.

According to a 2011 projection by the International Energy Agency, solar power generators may produce most of the world's electricity within 50 years, reducing the emissions of greenhouse gases that harm the environment. Cedric Philibert, senior analyst in the renewable energy division at the IEA said: "Photovoltaic and solar-thermal plants may meet most of the world's demand for electricity by 2060 – and half of all energy needs – with wind, hydropower and biomass plants supplying much of the remaining generation". "Photovoltaic and concentrated solar power together can become the major source of electricity", Philibert said.

In 2014 global wind power capacity expanded 16% to 369,553 MW. Yearly wind energy production is also growing rapidly and has reached around 4% of worldwide electricity usage, 11.4% in the EU, and it is widely used in Asia, and the United States. In 2015, worldwide installed photovoltaics capacity increased to 227 gigawatts (GW), sufficient to supply 1 percent of global electricity demands. Solar thermal energy stations operate in the USA and Spain, and as of 2016, the largest of these is the 392 MW Ivanpah Solar Electric Generating System in California. The world's largest geothermal power installation is The Geysers in California, with a rated capacity of 750 MW. Brazil has one of the largest renewable energy programs in the world, involving production of ethanol fuel from sugar cane, and ethanol now provides 18% of the country's automotive fuel. Ethanol fuel is also widely available in the USA.

Selected renewable energy global indicators	2008	2009	2010	2011	2012	2013	2014	2015
Investment in new renewable capacity (annual) (10⁹ USD)	182	178	237	279	256	232	270	285
Renewables power capacity (existing) (GWe)	1,140	1,230	1,320	1,360	1,470	1,578	1,712	1,849
Hydropower capacity (existing) (GWe)	885	915	945	970	990	1,018	1,055	1,064
Wind power capacity (existing) (GWe)	121	159	198	238	283	319	370	433
Solar PV capacity (grid-connected) (GWe)	16	23	40	70	100	138	177	227
Solar hot water capacity (existing) (GWth)	130	160	185	232	255	373	406	435
Ethanol production (annual) (10⁹ litres)	67	76	86	86	83	87	94	98
Biodiesel production (annual) (10⁹ litres)	12	17.8	18.5	21.4	22.5	26	29.7	30
Countries with policy targets for renewable energy use	79	89	98	118	138	144	164	173
Source: The Renewable Energy Policy Network for the 21st Century (REN21)–Global Status Report								

Economic Trends

Renewable energy technologies are getting cheaper, through technological change and through the benefits of mass production and market competition. A 2011 IEA report said: "A portfolio of renewable energy technologies is becoming cost-competitive in an increasingly broad range of circumstances, in some cases providing investment opportunities without the need for specific economic support," and added that "cost reductions in critical technologies, such as wind and solar, are set to continue."

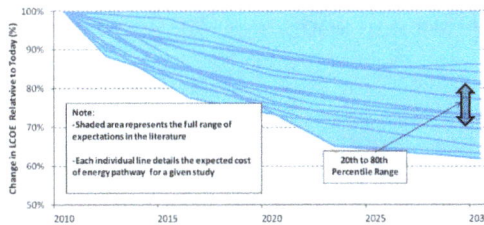

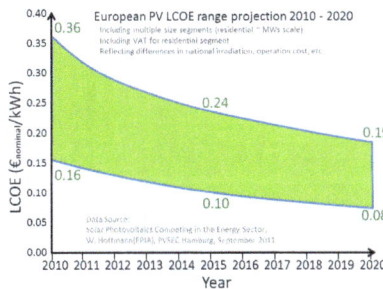

Projection of levelized cost for wind in the U.S. (left) and solar power in Europe

Hydro-electricity and geothermal electricity produced at favourable sites are now the cheapest way to generate electricity. Renewable energy costs continue to drop, and the levelised cost of electricity (LCOE) is declining for wind power, solar photovoltaic (PV), concentrated solar power (CSP) and some biomass technologies. Renewable energy is also the most economic solution for new grid-connected capacity in areas with good resources. As the cost of renewable power falls, the scope of economically viable applications increases. Renewable technologies are now often the most economic solution for new generating capacity. Where "oil-fired generation is the predominant power generation source (e.g. on islands, off-grid and in some countries) a lower-cost renewable solution almost always exists today". A series of studies by the US National Renewable Energy Laboratory modeled the "grid in the Western US under a number of different scenarios where intermittent renewables accounted for 33 percent of the total power." In the models, inefficiencies in cycling the fossil fuel plants to compensate for the variation in solar and wind energy resulted in an additional cost of "between $0.47 and $1.28 to each MegaWatt hour generated"; however, the savings in the cost of the fuels saved "adds up to $7 billion, meaning the added costs are, at most, two percent of the savings."

Hydroelectricity

Only 25% of the worlds estimated hydroelectric potential of 14,000 TWh/year has been developed, with Africa, Asia and Latin America having the greatest potential. The Three Gorges Dam in Hubei, China, has the world's largest instantaneous generating capacity (22,500 MW), with the Itaipu Dam in Brazil/Paraguay in second place (14,000 MW). The Three Gorges Dam is operated jointly with the much smaller Gezhouba Dam (3,115 MW). As of 2012, the total generating capacity of this two-dam complex is 25,615 MW. In 2008, this complex generated 98 TWh of electricity (81 TWh from the Three Gorges Dam and 17 TWh from the Gezhouba Dam), which is 3% more power in one year than the 95 TWh generated by Itaipu in 2008.

Wind Power Development

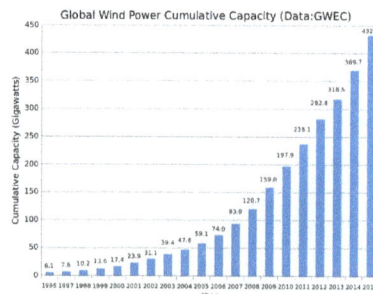

Worldwide growth of wind capacity (1996–2014)

Wind power is widely used in Europe, China, and the United States. From 2004 to 2014, worldwide installed capacity of wind power has been growing from 47 GW to

369 GW—a more than sevenfold increase within 10 years with 2014 breaking a new record in global installations (51 GW). As of the end of 2014, China, the United States and Germany combined accounted for half of total global capacity. Several other countries have achieved relatively high levels of wind power penetration, such as 21% of stationary electricity production in Denmark, 18% in Portugal, 16% in Spain, and 14% in Ireland in 2010 and have since continued to expand their installed capacity. More than 80 countries around the world are using wind power on a commercial basis.

Four offshore wind farms are in the Thames Estuary area: Kentish Flats, Gunfleet Sands, Thanet and London Array. The latter is the largest in the world as of April 2013.

- Offshore wind power

  As of 2014, offshore wind power amounted to 8,771 megawatt of global installed capacity. Although offshore capacity doubled within three years (from 4,117 MW in 2011), it accounted for only 2.3% of the total wind power capacity. The United Kingdom is the undisputed leader of offshore power with half of the world's installed capacity ahead of Denmark, Germany, Belgium and China.

- List of offshore and onshore wind farms

  As of 2012, the Alta Wind Energy Center (California, 1,020 MW) is the world's largest wind farm. The London Array (630 MW) is the largest offshore wind farm in the world. The United Kingdom is the world's leading generator of offshore wind power, followed by Denmark. There are several large offshore wind farms operational and under construction and these include Anholt (400 MW), BARD (400 MW), Clyde (548 MW), Fântânele-Cogealac (600 MW), Greater Gabbard (500 MW), Lincs (270 MW), London Array (630 MW), Lower Snake River (343 MW), Macarthur (420 MW), Shepherds Flat (845 MW), and the Sheringham Shoal (317 MW).

Solar Thermal

The United States conducted much early research in photovoltaics and concentrated solar power. The U.S. is among the top countries in the world in electricity generated by the Sun and several of the world's largest utility-scale installations are located in the desert Southwest.

The 377 MW Ivanpah Solar Electric Generating System with all three towers under load, Feb 2014. Taken from I-15.

The oldest solar thermal power plant in the world is the 354 megawatt (MW) SEGS thermal power plant, in California. The Ivanpah Solar Electric Generating System is a solar thermal power project in the California Mojave Desert, 40 miles (64 km) southwest of Las Vegas, with a gross capacity of 377 MW. The 280 MW Solana Generating Station is a solar power plant near Gila Bend, Arizona, about 70 miles (110 km) southwest of Phoenix, completed in 2013. When commissioned it was the largest parabolic trough plant in the world and the first U.S. solar plant with molten salt thermal energy storage.

Solar Towers of the PS10 and PS20 solar thermal plants in Spain

The solar thermal power industry is growing rapidly with 1.3 GW under construction in 2012 and more planned. Spain is the epicenter of solar thermal power development with 873 MW under construction, and a further 271 MW under development. In the United States, 5,600 MW of solar thermal power projects have been announced. Several power plants have been constructed in the Mojave Desert, Southwestern United States. The Ivanpah Solar Power Facility being the most recent. In developing countries, three World Bank projects for integrated solar thermal/combined-cycle gas-turbine power plants in Egypt, Mexico, and Morocco have been approved.

Photovoltaic Development

Photovoltaics (PV) uses solar cells assembled into solar panels to convert sunlight into electricity. It's a fast-growing technology doubling its worldwide installed capacity every couple of years. PV systems range from small, residential and commercial rooftop or building integrated installations, to large utility-scale photovoltaic power station. The predominant PV technology is crystalline silicon, while thin-film solar cell technology accounts for about 10 percent of global photovoltaic deployment. In recent years, PV technology has improved its electricity generating efficiency, reduced the installation cost per watt as well as its energy payback time, and has reached grid parity in at least 30 different markets by 2014. Financial institutions are predicting a second solar "gold rush" in the near future.

At the end of 2014, worldwide PV capacity reached at least 177,000 megawatts. Photovoltaics grew fastest in China, followed by Japan and the United States, while Germany remains the world's largest overall producer of photovoltaic power, contributing about 7.0 percent to the overall electricity generation. Italy meets 7.9 percent of its electricity demands with photovoltaic power—the highest share worldwide. For 2015, global cumulative capacity is forecasted to increase by more than 50 gigawatts (GW). By 2018, worldwide capacity is projected to reach as much as 430 gigawatts. This corresponds to a tripling within five years. Solar power is forecasted to become the world's largest source of electricity by 2050, with solar photovoltaics and concentrated solar power contributing 16% and 11%, respectively. This requires an increase of installed PV capacity to 4,600 GW, of which more than half is expected to be deployed in China and India.

Photovoltaic Power Stations

Commercial concentrated solar power plants were first developed in the 1980s. As the cost of solar electricity has fallen, the number of grid-connected solar PV systems has grown into the millions and utility-scale solar power stations with hundreds of megawatts are being built. Solar PV is rapidly becoming an inexpensive, low-carbon technology to harness renewable energy from the Sun.

Many solar photovoltaic power stations have been built, mainly in Europe, China and the USA. The 579 MW Solar Star, in the United States, is the world's largest PV power station.

Solar panels at the 550 MW Topaz Solar Farm

Many of these plants are integrated with agriculture and some use tracking systems that follow the sun's daily path across the sky to generate more electricity than fixed-mounted systems. There are no fuel costs or emissions during operation of the power stations.

Nellis Solar Power Plant, photovoltaic power plant in Nevada, USA

However, when it comes to renewable energy systems and PV, it is not just large systems that matter. Building-integrated photovoltaics or "onsite" PV systems use existing land and structures and generate power close to where it is consumed.

Biofuel Development

Biofuels provided 3% of the world's transport fuel in 2010. Mandates for blending biofuels exist in 31 countries at the national level and in 29 states/provinces. According to the International Energy Agency, biofuels have the potential to meet more than a quarter of world demand for transportation fuels by 2050.

Brazil produces bioethanol made from sugarcane available throughout the country. A typical gas station with dual fuel service is marked "A" for alcohol (ethanol) and "G" for gasoline.

Since the 1970s, Brazil has had an ethanol fuel program which has allowed the country to become the world's second largest producer of ethanol (after the United States) and the world's largest exporter. Brazil's ethanol fuel program uses modern equipment and cheap sugarcane as feedstock, and the residual cane-waste (bagasse) is used to produce heat and power. There are no longer light vehicles in Brazil running on pure gasoline. By the end of 2008 there were 35,000 filling stations throughout Brazil with at least one ethanol pump. Unfortunately, Operation Car Wash has seriously eroded public trust in oil companies and has implicated several high ranking Brazilian officials.

Nearly all the gasoline sold in the United States today is mixed with 10% ethanol, and motor vehicle manufacturers already produce vehicles designed to run on much higher ethanol blends. Ford, Daimler AG, and GM are among the automobile companies that sell "flexible-fuel" cars, trucks, and minivans that can use gasoline and ethanol blends ranging from pure gasoline up to 85% ethanol. By mid-2006, there were approximately 6 million ethanol compatible vehicles on U.S. roads.

Geothermal Development

Geothermal power is cost effective, reliable, sustainable, and environmentally friendly, but has historically been limited to areas near tectonic plate boundaries. Recent technological advances have expanded the range and size of viable resources, especially for applications such as home heating, opening a potential for widespread exploitation. Geothermal wells release greenhouse gases trapped deep within the earth, but these emissions are much lower per energy unit than those of fossil fuels. As a result, geothermal power has the potential to help mitigate global warming if widely deployed in place of fossil fuels.

Geothermal plant at The Geysers, California, USA

The International Geothermal Association (IGA) has reported that 10,715 MW of geothermal power in 24 countries is online, which is expected to generate 67,246 GWh of electricity in 2010. This represents a 20% increase in geothermal power online capacity since 2005. IGA projects this will grow to 18,500 MW by 2015, due to the large number of projects presently under consideration, often in areas previously assumed to have little exploitable resource.

In 2010, the United States led the world in geothermal electricity production with 3,086 MW of installed capacity from 77 power plants; the largest group of geothermal power plants in the world is located at The Geysers, a geothermal field in California. The Philippines follows the US as the second highest producer of geothermal power in the world, with 1,904 MW of capacity online; geothermal power makes up approximately 18% of the country's electricity generation.

Developing Countries

Renewable energy technology has sometimes been seen as a costly luxury item by critics, and affordable only in the affluent developed world. This erroneous view has persisted for many years, but 2015 was the first year when investment in non-hydro renewables, was higher in developing countries, with $156 billion invested, mainly in China, India, and Brazil.

Solar cookers use sunlight as energy source for outdoor cooking

Renewable energy can be particularly suitable for developing countries. In rural and remote areas, transmission and distribution of energy generated from fossil fuels can be difficult and expensive. Producing renewable energy locally can offer a viable alternative.

Technology advances are opening up a huge new market for solar power: the approximately 1.3 billion people around the world who don't have access to grid electricity. Even though they are typically very poor, these people have to pay far more for lighting than people in rich countries because they use inefficient kerosene lamps. Solar power costs half as much as lighting with kerosene. As of 2010, an estimated 3 million households get power from small solar PV systems. Kenya is the world leader in the number of solar power systems installed per capita. More than 30,000 very small solar panels, each producing 12 to 30 watts, are sold in Kenya annually. Some Small Island Developing States (SIDS) are also turning to solar power to reduce their costs and increase their sustainability.

Micro-hydro configured into mini-grids also provide power. Over 44 million households use biogas made in household-scale digesters for lighting and/or cooking, and more than 166 million households rely on a new generation of more-efficient biomass cookstoves. Clean liquid fuel sourced from renewable feedstocks are used for cooking and lighting in energy-poor areas of the developing world. Alcohol fuels (ethanol and methanol) can be produced sustainably from non-food sugary, starchy, and cellulostic feedstocks. Project Gaia, Inc. and CleanStar Mozambique are implementing clean cooking programs with liquid ethanol stoves in Ethiopia, Kenya, Nigeria and Mozambique.

Renewable energy projects in many developing countries have demonstrated that renewable energy can directly contribute to poverty reduction by providing the energy needed for creating businesses and employment. Renewable energy technologies can also make indirect contributions to alleviating poverty by providing energy for cooking, space heating, and lighting. Renewable energy can also contribute to education, by providing electricity to schools.

Industry and Policy Trends

U.S. President Barack Obama's American Recovery and Reinvestment Act of 2009 includes more than $70 billion in direct spending and tax credits for clean energy and associated transportation programs. Leading renewable energy companies include First Solar, Gamesa, GE Energy, Hanwha Q Cells, Sharp Solar, Siemens, SunOpta, Suntech Power, and Vestas.

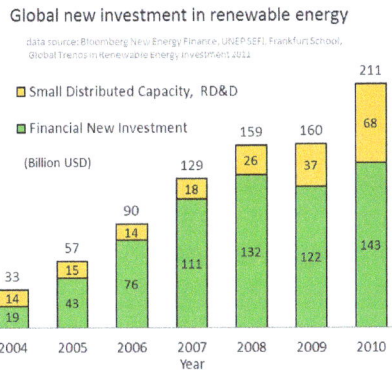

Global New Investments in Renewable Energy

Many national, state, and local governments have also created green banks. A green bank is a quasi-public financial institution that uses public capital to leverage private investment in clean energy technologies. Green banks use a variety of financial tools to bridge market gaps that hinder the deployment of clean energy.

The military has also focused on the use of renewable fuels for military vehicles. Unlike fossil fuels, renewable fuels can be produced in any country, creating a strategic advantage. The US military has already committed itself to have 50% of its energy consumption come from alternative sources.

The International Renewable Energy Agency (IRENA) is an intergovernmental organization for promoting the adoption of renewable energy worldwide. It aims to provide concrete policy advice and facilitate capacity building and technology transfer. IRENA was formed on 26 January 2009, by 75 countries signing the charter of IRENA. As of March 2010, IRENA has 143 member states who all are considered as founding members, of which 14 have also ratified the statute.

As of 2011, 119 countries have some form of national renewable energy policy target or renewable support policy. National targets now exist in at least 98 countries. There is also a wide range of policies at state/provincial and local levels.

United Nations' Secretary-General Ban Ki-moon has said that renewable energy has the ability to lift the poorest nations to new levels of prosperity. In October 2011, he "announced the creation of a high-level group to drum up support for energy access, energy efficiency and greater use of renewable energy. The group is to be co-chaired by Kandeh Yumkella, the chair of UN Energy and director general of the UN Industrial Development Organisation, and Charles Holliday, chairman of Bank of America".

100% Renewable Energy

The incentive to use 100% renewable energy, for electricity, transport, or even total primary energy supply globally, has been motivated by global warming and other ecological as well as economic concerns. The Intergovernmental Panel on Climate Change has said that there are few fundamental technological limits to integrating a portfolio of renewable energy technologies to meet most of total global energy demand. Renewable energy use has grown much faster than even advocates anticipated. At the national level, at least 30 nations around the world already have renewable energy contributing more than 20% of energy supply. Also, Professors S. Pacala and Robert H. Socolow have developed a series of "stabilization wedges" that can allow us to maintain our quality of life while avoiding catastrophic climate change, and "renewable energy sources," in aggregate, constitute the largest number of their "wedges."

Using 100% renewable energy was first suggested in a Science paper published in 1975 by Danish physicist Bent Sørensen. It was followed by several other proposals, until in 1998 the first detailed analysis of scenarios with very high shares of renewables were published. These were followed by the first detailed 100% scenarios. In 2006 a PhD thesis was published by Czisch in which it was shown that in a 100% renewable scenario energy supply could match demand in every hour of the year in Europa and North Africa. In the same year Danish Energy professor Henrik Lund published a first paper in which he addresses the optimal combination of renewables, which was followed by several other papers on the transition to 100% renewable energy in Denmark. Since then Lund has been publishing several papers on 100% renewable energy. After 2009 publications began to rise steeply, covering 100% scenarios for countries in Europa, America, Australia and other parts of the world.

In 2011 Mark Z. Jacobson, professor of civil and environmental engineering at Stanford University, and Mark Delucchi published a study on 100% renewable global energy supply in the journal Energy Policy. They found producing all new energy with wind power, solar power, and hydropower by 2030 is feasible and existing energy supply arrangements could be replaced by 2050. Barriers to implementing the renewable energy plan are seen to be "primarily social and political, not technological or economic". They also found that energy costs with a wind, solar, water system should be similar to today's energy costs.

Similarly, in the United States, the independent National Research Council has noted that "sufficient domestic renewable resources exist to allow renewable electricity to play a significant role in future electricity generation and thus help confront issues related to climate change, energy security, and the escalation of energy costs ... Renewable energy is an attractive option because renewable resources available in the United States, taken collectively, can supply significantly greater amounts of electricity than the total current or projected domestic demand." .

The most significant barriers to the widespread implementation of large-scale renewable energy and low carbon energy strategies are primarily political and not technological. According to the 2013 *Post Carbon Pathways* report, which reviewed many international studies, the key roadblocks are: climate change denial, the fossil fuels lobby, political inaction, unsustainable energy consumption, outdated energy infrastructure, and financial constraints.

Emerging Technologies

Other renewable energy technologies are still under development, and include cellulosic ethanol, hot-dry-rock geothermal power, and marine energy. These technologies are not yet widely demonstrated or have limited commercialization. Many are on the horizon and may have potential comparable to other renewable energy technologies, but still depend on attracting sufficient attention and research, development and demonstration (RD&D) funding.

There are numerous organizations within the academic, federal, and commercial sectors conducting large scale advanced research in the field of renewable energy. This research spans several areas of focus across the renewable energy spectrum. Most of the research is targeted at improving efficiency and increasing overall energy yields. Multiple federally supported research organizations have focused on renewable energy in recent years. Two of the most prominent of these labs are Sandia National Laboratories and the National Renewable Energy Laboratory (NREL), both of which are funded by the United States Department of Energy and supported by various corporate partners. Sandia has a total budget of $2.4 billion while NREL has a budget of $375 million.

- Enhanced geothermal system

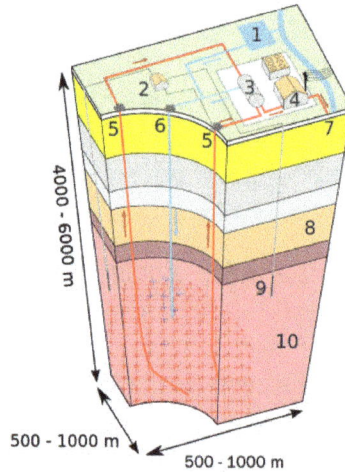

Enhanced geothermal system

Enhanced geothermal systems (EGS) are a new type of geothermal power technologies that do not require natural convective hydrothermal resources. The vast majority of geothermal energy within drilling reach is in dry and non-porous rock. EGS technologies "enhance" and/or create geothermal resources in this "hot dry rock (HDR)" through hydraulic stimulation. EGS and HDR technologies, like hydrothermal geothermal, are expected to be baseload resources which produce power 24 hours a day like a fossil plant. Distinct from hydrothermal, HDR and EGS may be feasible anywhere in the world, depending on the economic limits of drill depth. Good locations are over deep granite covered by a thick (3–5 km) layer of insulating sediments which slow heat loss. There are HDR and EGS systems currently being developed and tested in France, Australia, Japan, Germany, the U.S. and Switzerland. The largest EGS project in the world is a 25 megawatt demonstration plant currently being developed in the Cooper Basin, Australia. The Cooper Basin has the potential to generate 5,000–10,000 MW.

- Cellulosic Ethanol

Several refineries that can process biomass and turn it into ethano are built by companies such as Iogen, POET, and Abengoa, while other companies such as the Verenium Corporation, Novozymes, and Dyadic International are producing enzymes which could enable future commercialization. The shift from food crop feedstocks to waste residues and native grasses offers significant opportunities for a range of players, from farmers to biotechnology firms, and from project developers to investors.

- Marine Energy

Marine energy (also sometimes referred to as ocean energy) refers to the energy carried by ocean waves, tides, salinity, and ocean temperature differences. The movement of water in the world's oceans creates a vast store of kinetic energy, or energy in motion. This energy can be harnessed to generate electricity to power homes, transport and industries. The term marine energy encompasses both wave power – power from surface waves, and tidal power – obtained from the kinetic energy of large bodies of moving water. Reverse electrodialysis (RED) is a technology for generating electricity by mixing fresh river water and salty sea water in large power cells designed for this purpose; as of 2016 it is being tested at a small scale (50 kW). Offshore wind power is not a form of marine energy, as wind power is derived from the wind, even if the wind turbines are placed over water. The oceans have a tremendous amount of energy and are close to many if not most concentrated populations. Ocean energy has the potential of providing a substantial amount of new renewable energy around the world.

Rance Tidal Power Station, France

#	Station	Country	Location	Capacity
1.	Sihwa Lake Tidal Power Station	South Korea	37°18′47″N 126°36′46″E / 37.31306°N 126.61278°E / 37.31306; 126.61278 (Sihwa Lake Tidal Power Station)	254 MW
2.	Rance Tidal Power Station	France	48°37′05″N 02°01′24″W / 48.61806°N 2.02333°W / 48.61806; -2.02333 (Rance Tidal Power Station)	240 MW
3.	Annapolis Royal Generating Station	Canada	44°45′07″N 65°30′40″W / 44.75194°N 65.51111°W / 44.75194; -65.51111 (Annapolis Royal Generating Station)	20 MW

- Experimental Solar Power

Concentrated photovoltaics (CPV) systems employ sunlight concentrated onto photovoltaic surfaces for the purpose of electricity generation. Thermoelectric, or "thermovoltaic" devices convert a temperature difference between dissimilar materials into an electric current.

- Floating Solar Arrays

Floating solar arrays are PV systems that float on the surface of drinking water reservoirs, quarry lakes, irrigation canals or remediation and tailing ponds. A small number of such systems exist in France, India, Japan, South Korea, the United Kingdom, Singapore and the United States. The systems are said to have advantages over photovoltaics on land. The cost of land is more expensive, and there are fewer rules and regulations for structures built on bodies of water not used for recreation. Unlike most land-based solar plants, floating arrays can be unobtrusive because they are hidden from public view. They achieve higher efficiencies than PV panels on land, because water cools the panels. The panels have a special coating to prevent rust or corrosion. In May 2008, the Far Niente Winery in Oakville, California, pioneered the world's first floatovoltaic system by installing 994 solar PV modules with a total capacity of 477 kW onto 130 pontoons and floating them on the winery's irrigation pond. Utility-scale floating PV farms are starting to be built. Kyocera will develop the world's largest, a 13.4 MW farm on the reservoir above Yamakura Dam in Chiba Prefecture using 50,000 solar panels. Salt-water resistant floating farms are also being constructed for ocean use. The largest so far announced floatovoltaic project is a 350 MW power station in the Amazon region of Brazil.

- Solar-assisted Heat Pump

A heat pump is a device that provides heat energy from a source of heat to a destination called a "heat sink". Heat pumps are designed to move thermal energy opposite to the direction of spontaneous heat flow by absorbing heat from a cold space and releasing it to a warmer one. A solar-assisted heat pump represents the integration of a heat pump and thermal solar panels in a single integrated system. Typically these two technologies are used separately (or only placing them in parallel) to produce hot water. In this system the solar thermal panel performs the function of the low temperature heat source and the heat produced is used to feed the heat pump's evaporator. The goal of this system is to get high COP and then produce energy in a more efficient and less expensive way.

It is possible to use any type of solar thermal panel (sheet and tubes, roll-bond, heat pipe, thermal plates) or hybrid (mono/polycrystalline, thin film) in combination with the heat pump. The use of a hybrid panel is preferable because it allows to cover a part of the electricity demand of the heat pump and reduce the power consumption and consequently the variable costs of the system.

- Artificial Photosynthesis

Artificial photosynthesis uses techniques including nanotechnology to store solar electromagnetic energy in chemical bonds by splitting water to produce hydrogen and then using carbon dioxide to make methanol. Researchers in this field are striving to design molecular mimics of photosynthesis that utilize a wider region of the solar spectrum, employ catalytic systems made from abundant, inexpensive materials that are robust, readily repaired, non-toxic, stable in a variety of environmental conditions and perform more efficiently allowing a greater proportion of photon energy to end up in the storage compounds, i.e., carbohydrates (rather than building and sustaining living cells). However, prominent research faces hurdles, Sun Catalytix a MIT spin-off stopped scaling up their prototype fuel-cell in 2012, because it offers few savings over other ways to make hydrogen from sunlight.

- Algae Fuels

Producing liquid fuels from oil-rich varieties of algae is an ongoing research topic. Various microalgae grown in open or closed systems are being tried including some system that can be set up in brownfield and desert lands.

- Space-based Solar Power

For either photovoltaic or thermal systems, one option is to loft them into space, particularly Geosynchronous orbit. To be competitive with Earth-based solar power systems, the specific mass (kg/kW) times the cost to loft mass plus the cost of the parts needs to be $2400 or less. I.e., for a parts cost plus rectenna of $1100/kW, the product of the $/kg and kg/kW must be $1300/kW or less. Thus for 6.5 kg/kW, the transport cost cannot exceed $200/kg. While that will require a 100 to one reduction, SpaceX is targeting a ten to one reduction, Reaction Engines may make a 100 to one reduction possible.

- Carbon-neutral and Negative Fuels

Carbon-neutral fuels are synthetic fuels (including methane, gasoline, diesel fuel, jet fuel or ammonia) produced by hydrogenating waste carbon dioxide recycled from power plant flue-gas emissions, recovered from automotive exhaust gas, or derived from carbonic acid in seawater. Such fuels are considered carbon-neutral because they do not result in a net increase in atmospheric greenhouse gases. T

Carbon-neutral fuels offer relatively low cost energy storage, alleviating the problems of wind and solar variability, and they enable distribution of wind, water, and solar power through existing natural gas pipelines. Nighttime wind power is considered the most economical form of electrical power with which to synthesize fuel, because the load curve for electricity peaks sharply during the warmest hours of the day, but wind tends to blow slightly more at night than during the day, so, the price of nighttime wind power is often much less expensive than any alternative. Germany has built a 250 kilowatt synthetic methane plant which they are scaling up to 10 megawatts.

The George Olah carbon dioxide recycling plant in Grindavík, Iceland has been producing 2 million liters of methanol transportation fuel per year from flue exhaust of the Svartsengi Power Station since 2011. It has the capacity to produce 5 million liters per year.

Debate

Renewable electricity production, from sources such as wind power and solar power, is sometimes criticized for being variable or intermittent, but is not true for concentrated solar, geothermal and biofuels, that have continuity. In any case, the International Energy Agency has stated that deployment of renewable technologies usually increases the diversity of electricity sources and, through local generation, contributes to the flexibility of the system and its resistance to central shocks.

There have been "not in my back yard" (NIMBY) concerns relating to the visual and other impacts of some wind farms, with local residents sometimes fighting or blocking construction. In the USA, the Massachusetts Cape Wind project was delayed for years partly because of aesthetic concerns. However, residents in other areas have been more positive. According to a town councilor, the overwhelming majority of locals believe that the Ardrossan Wind Farm in Scotland has enhanced the area.

A recent UK Government document states that "projects are generally more likely to succeed if they have broad public support and the consent of local communities. This means giving communities both a say and a stake". In countries such as Germany and Denmark many renewable projects are owned by communities, particularly through cooperative structures, and contribute significantly to overall levels of renewable energy deployment.

The market for renewable energy technologies has continued to grow. Climate change concerns and increasing in green jobs, coupled with high oil prices, peak oil, oil wars, oil spills, promotion of electric vehicles and renewable electricity, nuclear disasters and increasing government support, are driving increasing renewable energy legislation, incentives and commercialization. New government spending, regulation and policies helped the industry weather the 2009 economic crisis better than many other sectors.

Trends and Technology in Renewable Energy

Wind Power

Wind power is the use of air flow through wind turbines to mechanically power generators for electricity. Wind power, as an alternative to burning fossil fuels, is plentiful, renewable, widely distributed, clean, produces no greenhouse gas emissions during operation, uses no water, and uses little land. The net effects on the environment are far less problematic than those of nonrenewable power sources.

Wind farms consist of many individual wind turbines which are connected to the electric power transmission network. Onshore wind is an inexpensive source of electricity, competitive with or in many places cheaper than coal or gas plants. Offshore wind is steadier and stronger than on land, and offshore farms have less visual impact, but construction and maintenance costs are considerably higher. Small onshore wind farms can feed some energy into the grid or provide electricity to isolated off-grid locations.

Wind power stations in Xinjiang, China

Wind power gives variable power which is very consistent from year to year but which has significant variation over shorter time scales. It is therefore used in conjunction with other electric power sources to give a reliable supply. As the proportion of wind power in a region increases, a need to upgrade the grid, and a lowered ability to supplant conventional production can occur. Power management techniques such as having excess capacity, geographically distributed turbines, dispatchable backing sources, sufficient hydroelectric power, exporting and importing power to neighboring areas, using vehicle-to-grid strategies or reducing demand when wind production is low, can in many cases overcome these problems. In addition, weather forecasting permits the electricity network to be readied for the predictable variations in production that occur.

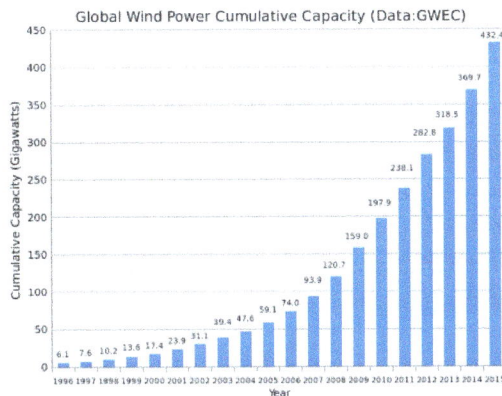

Global growth of installed capacity

As of 2015, Denmark generates 40% of its electricity from wind, and at least 83 other countries around the world are using wind power to supply their electricity grids. In 2014 global wind power capacity expanded 16% to 369,553 MW. Yearly wind energy production is also growing rapidly and has reached around 4% of worldwide electricity usage, 11.4% in the EU.

History

Wind power has been used as long as humans have put sails into the wind. For more than two millennia wind-powered machines have ground grain and pumped water. Wind power was widely available and not confined to the banks of fast-flowing streams, or later, requiring sources of fuel. Wind-powered pumps drained the polders of the Netherlands, and in arid regions such as the American mid-west or the Australian outback, wind pumps provided water for live stock and steam engines.

Charles Brush's windmill of 1888, used for generating electricity.

The first windmill used for the production of electricity was built in Scotland in July 1887 by Prof James Blyth of Anderson's College, Glasgow (the precursor of Strathclyde University). Blyth's 10 m high, cloth-sailed wind turbine was installed in the garden of his holiday cottage at Marykirk in Kincardineshire and was used to charge accumulators developed by the Frenchman Camille Alphonse Faure, to power the lighting in the cottage, thus making it the first house in the world to have its electricity supplied by wind power. Blyth offered the surplus electricity to the people of Marykirk for lighting the main street, however, they turned down the offer as they thought electricity was "the work of the devil." Although he later built a wind turbine to supply emergency power to the local Lunatic Asylum, Infirmary and Dispensary of Montrose the invention never really caught on as the technology was not considered to be economically viable.

Across the Atlantic, in Cleveland, Ohio a larger and heavily engineered machine was designed and constructed in the winter of 1887–1888 by Charles F. Brush, this was built by his engineering company at his home and operated from 1886 until 1900. The Brush wind turbine had a rotor 17 m (56 foot) in diameter and was mounted on an 18 m (60 foot) tower. Although large by today's standards, the machine was only rated at 12 kW. The connected dynamo was used either to charge a bank of batteries or to operate up to 100 incandescent light bulbs, three arc lamps, and various motors in Brush's laboratory.

With the development of electric power, wind power found new applications in lighting buildings remote from centrally-generated power. Throughout the 20th century parallel paths developed small wind stations suitable for farms or residences, and larger utility-scale wind generators that could be connected to electricity grids for remote use of power. Today wind powered generators operate in every size range between tiny stations for battery charging at isolated residences, up to near-gigawatt sized offshore wind farms that provide electricity to national electrical networks.

A wind farm is a group of wind turbines in the same location used for production of electricity. A large wind farm may consist of several hundred individual wind turbines distributed over an extended area, but the land between the turbines may be used for agricultural or other purposes. For example, Gansu Wind Farm, the largest wind farm in the world, has several thousand turbines. A wind farm may also be located offshore.

Almost all large wind turbines have the same design — a horizontal axis wind turbine having an upwind rotor with three blades, attached to a nacelle on top of a tall tubular tower.

In a wind farm, individual turbines are interconnected with a medium voltage (often 34.5 kV), power collection system and communications network. In general, a distance of 7D (7 × Rotor Diameter of the Wind Turbine) is set between each turbine in a fully developed wind farm. At a substation, this medium-voltage electric current is increased in voltage with a transformer for connection to the high voltage electric power transmission system.

Generator Characteristics and Stability

Induction generators, which were often used for wind power projects in the 1980s and 1990s, require reactive power for excitation so substations used in wind-power collection systems include substantial capacitor banks for power factor correction. Different types of wind turbine generators behave differently during transmission grid disturbances, so extensive modelling of the dynamic electromechanical characteristics of a new wind farm is required by transmission system operators to ensure predictable stable behaviour during system faults. In particular, induction generators cannot support the system voltage during faults, unlike steam or hydro turbine-driven synchronous generators.

Today these generators aren't used any more in modern turbines. Instead today most turbines use variable speed generators combined with partial- or full-scale power converter between the turbine generator and the collector system, which generally have more desirable properties for grid interconnection and have Low voltage ride through-capabilities. Modern concepts use either doubly fed machines with partial-scale converters or squirrel-cage induction generators or synchronous generators (both permanently and electrically excited) with full scale converters.

Transmission systems operators will supply a wind farm developer with a grid code to specify the requirements for interconnection to the transmission grid. This will include power factor, constancy of frequency and dynamic behaviour of the wind farm turbines during a system fault.

Offshore Wind Power

Offshore wind power refers to the construction of wind farms in large bodies of water to generate electricity. These installations can utilize the more frequent and powerful winds that are available in these locations and have less aesthetic impact on the landscape than land based projects. However, the construction and the maintenance costs are considerably higher.

The world's second full-scale floating wind turbine (and first to be installed without the use of heavy-lift vessels), WindFloat, operating at rated capacity (2 MW) approximately 5 km offshore of Póvoa de Varzim, Portugal

Siemens and Vestas are the leading turbine suppliers for offshore wind power. DONG Energy, Vattenfall and E.ON are the leading offshore operators. As of October 2010, 3.16 GW of offshore wind power capacity was operational, mainly in Northern Europe. According to BTM Consult, more than 16 GW of additional capacity will be installed before the end of 2014 and the UK and Germany will become the two leading markets. Offshore wind power capacity is expected to reach a total of 75 GW worldwide by 2020, with significant contributions from China and the US.

In 2012, 1,662 turbines at 55 offshore wind farms in 10 European countries produced 18 TWh, enough to power almost five million households. As of August 2013 the London Array in the United Kingdom is the largest offshore wind farm in the world at 630 MW. This is followed by Gwynt y Môr (576 MW), also in the UK.

World's largest offshore wind farms				
Wind farm	**Capacity (MW)**	**Country**	**Turbines and model**	**Commissioned**
London Array	630	🇬🇧 United Kingdom	175 × Siemens SWT-3.6	2012
Gwynt y Môr	576	🇬🇧 United Kingdom	160 × Siemens SWT-3.6 107	2015
Greater Gabbard	504	🇬🇧 United Kingdom	140 × Siemens SWT-3.6	2012
Anholt	400	🇩🇰 Denmark	111 × Siemens SWT-3.6–120	2013
BARD Offshore 1	400	🇩🇪 Germany	80 BARD 5.0 turbines	2013

Collection and Transmission Network

In a wind farm, individual turbines are interconnected with a medium voltage (usually 34.5 kV) power collection system and communications network. At a substation, this medium-voltage electric current is increased in voltage with a transformer for connection to the high voltage electric power transmission system.

A transmission line is required to bring the generated power to (often remote) markets. For an off-shore station this may require a submarine cable. Construction of a new high-voltage line may be too costly for the wind resource alone, but wind sites may take advantage of lines installed for conventionally fueled generation.

One of the biggest current challenges to wind power grid integration in the United States is the necessity of developing new transmission lines to carry power from wind farms, usually in remote lowly populated states in the middle of the country due to availability of wind, to high load locations, usually on the coasts where population density is higher. The current transmission lines in remote locations were not designed for the transport of large amounts of energy. As transmission lines become longer the losses associated with power transmission increase, as modes of losses at lower lengths are exacerbated and new modes of losses are no longer negligible as the length is increased, making it harder to transport large loads over large distances. However, resistance from state and local governments makes it difficult to construct new transmission lines. Multi state power transmission projects are discouraged by states with cheap electricity rates for fear that exporting their cheap power will lead to increased rates. A 2005 energy law gave the Energy Department authority to approve transmission projects states refused to act on, but after an attempt to use this authority, the Senate declared the department was being overly aggressive in doing so. Another problem is that wind companies find out after the fact that the transmission capacity of a new farm is below the generation capacity, largely because federal utility rules to encourage renewable energy installation allow feeder lines to meet only minimum standards. These are important issues that need to be solved, as when the transmission capacity does not meet the generation capacity, wind farms are forced to produce below their full potential or stop running all together, in a process known as curtailment. While this leads to potential renewable generation left untapped, it prevents possible grid overload or risk to reliable service.

Wind Power Capacity and Production

Worldwide there are now over two hundred thousand wind turbines operating, with a total nameplate capacity of 432,000 MW as of end 2015. The European Union alone passed some 100,000 MW nameplate capacity in September 2012, while the United States surpassed 75,000 MW in 2015 and China's grid connected capacity passed 145,000 MW in 2015.

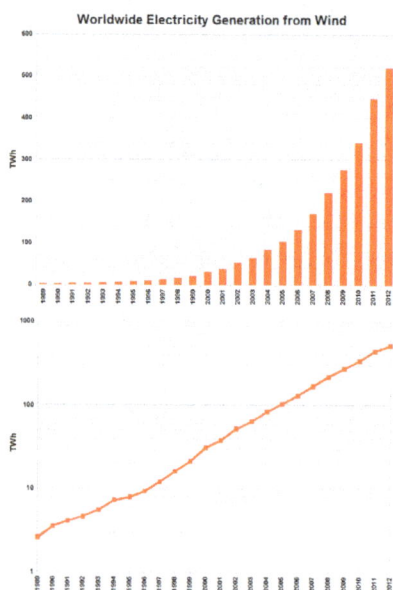

World wind generation capacity more than quadrupled between 2000 and 2006, doubling about every three years. The United States pioneered wind farms and led the world in installed capacity in the 1980s and into the 1990s. In 1997 installed capacity in Germany surpassed the U.S. and led until once again overtaken by the U.S. in 2008. China has been rapidly expanding its wind installations in the late 2000s and passed the U.S. in 2010 to become the world leader. As of 2011, 83 countries around the world were using wind power on a commercial basis.

Wind power capacity has expanded rapidly to 336 GW in June 2014, and wind energy production was around 4% of total worldwide electricity usage, and growing rapidly. The actual amount of electricity that wind is able to generate is calculated by multiplying the nameplate capacity by the capacity factor, which varies according to equipment and location. Estimates of the capacity factors for wind installations are in the range of 35% to 44%.

Europe accounted for 48% of the world total wind power generation capacity in 2009. In 2010, Spain became Europe's leading producer of wind energy, achieving 42,976 GWh. Germany held the top spot in Europe in terms of installed capacity, with a total of 27,215 MW as of 31 December 2010. In 2015 wind power constituted 15.6% of all installed power generation capacity in the EU and it generates around 11.4% of its power.

Top windpower electricity producing countries in 2012 (TWh)		
Country	**Windpower Production**	**% of World Total**
United States	140.9	26.4
China	118.1	22.1
Spain	49.1	9.2
Germany	46.0	8.6
India	30.0	5.6
United Kingdom	19.6	3.7
France	14.9	2.8
Italy	13.4	2.5
Canada	11.8	2.2
Denmark	10.3	1.9
(rest of world)	80.2	15.0
World Total	**534.3 TWh**	**100%**
Source:*Observ'ER – Electricity Production From Wind Sources [2012]*		

Growth Trends

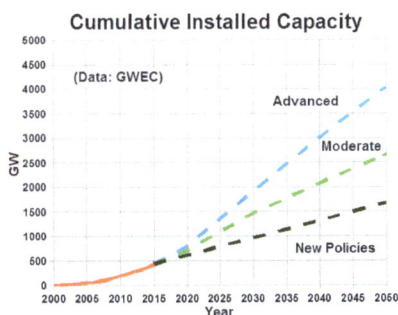

Worldwide installed wind power capacity forecast (Source: Global Wind Energy Council)

After setting new records in 2014, the wind power industry surprised many observers with another record breaking year in 2015, chalking up 22% annual market growth and passing the 60 GW mark for the first time in a single year; and this after having broken the 50 GW mark for the first time in 2014. In 2015, close to half of all new wind power was added outside of the traditional markets in Europe and North America. This was largely from new construction in China and India. Global Wind Energy Council (GWEC) figures show that 2015 recorded an increase of installed capacity of more than

63 GW, taking the total installed wind energy capacity to 432.9 GW, up from 74 GW in 2006. In terms of economic value, the wind energy sector has become one of the important players in the energy markets, with the total investments reaching US$329bn (€296.6bn), an increase of 4% over 2014.

Although the wind power industry was affected by the global financial crisis in 2009 and 2010, GWEC predicts that the installed capacity of wind power will be 792.1 GW by the end of 2020 and 4,042 GW by end of 2050. The increased commissioning of wind power is being accompanied by record low prices for forthcoming renewable electricity. In some cases, wind onshore is already the cheapest electricity generation option and costs are continuing to decline. The contracted prices for wind onshore for the next few years are now as low as 30 USD/MWh.

In the EU in 2015, 44% of all new generating capacity was wind power; while in the same period net fossil fuel power capacity decreased.

Capacity Factor

Since wind speed is not constant, a wind farm's annual energy production is never as much as the sum of the generator nameplate ratings multiplied by the total hours in a year. The ratio of actual productivity in a year to this theoretical maximum is called the capacity factor. Typical capacity factors are 15–50%; values at the upper end of the range are achieved in favourable sites and are due to wind turbine design improvements.

Online data is available for some locations, and the capacity factor can be calculated from the yearly output. For example, the German nationwide average wind power capacity factor over all of 2012 was just under 17.5% (45867 GW·h/yr / (29.9 GW × 24 × 366) = 0.1746), and the capacity factor for Scottish wind farms averaged 24% between 2008 and 2010.

Unlike fueled generating plants, the capacity factor is affected by several parameters, including the variability of the wind at the site and the size of the generator relative to the turbine's swept area. A small generator would be cheaper and achieve a higher capacity factor but would produce less electricity (and thus less profit) in high winds. Conversely, a large generator would cost more but generate little extra power and, depending on the type, may stall out at low wind speed. Thus an optimum capacity factor of around 40–50% would be aimed for.

A 2008 study released by the U.S. Department of Energy noted that the capacity factor of new wind installations was increasing as the technology improves, and projected further improvements for future capacity factors. In 2010, the department estimated the capacity factor of new wind turbines in 2010 to be 45%. The annual average capacity factor for wind generation in the US has varied between 29.8% and 34.0% during the period 2010–2015.

Penetration

Wind energy penetration is the fraction of energy produced by wind compared with the total generation. There is no generally accepted maximum level of wind penetration. The limit for a particular grid will depend on the existing generating plants, pricing mechanisms, capacity for energy storage, demand management and other factors. An interconnected electricity grid will already include reserve generating and transmission capacity to allow for equipment failures. This reserve capacity can also serve to compensate for the varying power generation produced by wind stations. Studies have indicated that 20% of the total annual electrical energy consumption may be incorporated with minimal difficulty. These studies have been for locations with geographically dispersed wind farms, some degree of dispatchable energy or hydropower with storage capacity, demand management, and interconnected to a large grid area enabling the export of electricity when needed. Beyond the 20% level, there are few technical limits, but the economic implications become more significant. Electrical utilities continue to study the effects of large scale penetration of wind generation on system stability and economics.

Country	Penetration
Denmark (2015)	42.1%
Portugal (2011)	19%
Spain (2011)	16%
Ireland (2012)	16%
United Kingdom (2014)	9.3%
Germany (2011)	8%
United States (2013)	4.5%

A wind energy penetration figure can be specified for different durations of time, but is often quoted annually. To obtain 100% from wind annually requires substantial long term storage or substantial interconnection to other systems which may already have substantial storage. On a monthly, weekly, daily, or hourly basis—or less—wind might supply as much as or more than 100% of current use, with the rest stored or exported. Seasonal industry might then take advantage of high wind and low usage times such as at night when wind output can exceed normal demand. Such industry might include production of silicon, aluminum, steel, or of natural gas, and hydrogen, and using future long term storage to facilitate 100% energy from variable renewable energy. Homes can also be programmed to accept extra electricity on demand, for example by remotely turning up water heater thermostats.

In Australia, the state of South Australia generates around half of the nation's wind power capacity. By the end of 2011 wind power in South Australia, championed by Premier (and Climate Change Minister) Mike Rann, reached 26% of the State's electricity generation, edging out coal for the first time. At this stage South Australia, with only 7.2% of Australia's population, had 54% of Australia's installed capacity.

Variability

Electricity generated from wind power can be highly variable at several different timescales: hourly, daily, or seasonally. Annual variation also exists, but is not as significant. Because instantaneous electrical generation and consumption must remain in balance to maintain grid stability, this variability can present substantial challenges to incorporating large amounts of wind power into a grid system. Intermittency and the non-dispatchable nature of wind energy production can raise costs for regulation, incremental operating reserve, and (at high penetration levels) could require an increase in the already existing energy demand management, load shedding, storage solutions or system interconnection with HVDC cables.

Windmills are typically installed in favourable windy locations. In the image, wind power generators in Spain, near an Osborne bull.

Fluctuations in load and allowance for failure of large fossil-fuel generating units require reserve capacity that can also compensate for variability of wind generation.

Increase in system operation costs, Euros per MWh, for 10% & 20% wind share		
Country	**10%**	**20%**
Germany	2.5	3.2
Denmark	0.4	0.8
Finland	0.3	1.5
Norway	0.1	0.3
Sweden	0.3	0.7

Wind power is variable, and during low wind periods it must be replaced by other power sources. Transmission networks presently cope with outages of other generation plants and daily changes in electrical demand, but the variability of intermittent power sources such as wind power, are unlike those of conventional power generation plants which, when scheduled to be operating, may be able to deliver their nameplate capacity around 95% of the time.

Presently, grid systems with large wind penetration require a small increase in the frequency of usage of natural gas spinning reserve power plants to prevent a loss of electricity in the event that conditions are not favorable for power production from the wind. At lower wind power grid penetration, this is less of an issue.

GE has installed a prototype wind turbine with onboard battery similar to that of an electric car, equivalent of 1 minute of production. Despite the small capacity, it is enough to guarantee that power output complies with forecast for 15 minutes, as the battery is used to eliminate the difference rather than provide full output. The increased predictability can be used to take wind power penetration from 20 to 30 or 40 per cent. The battery cost can be retrieved by selling burst power on demand and reducing backup needs from gas plants.

A report on Denmark's wind power noted that their wind power network provided less than 1% of average demand on 54 days during the year 2002. Wind power advocates argue that these periods of low wind can be dealt with by simply restarting existing power stations that have been held in readiness, or interlinking with HVDC. Electrical grids with slow-responding thermal power plants and without ties to networks with hydroelectric generation may have to limit the use of wind power. According to a 2007 Stanford University study published in the *Journal of Applied Meteorology and Climatology*, interconnecting ten or more wind farms can allow an average of 33% of the total energy produced (i.e. about 8% of total nameplate capacity) to be used as reliable, baseload electric power which can be relied on to handle peak loads, as long as minimum criteria are met for wind speed and turbine height.

Conversely, on particularly windy days, even with penetration levels of 16%, wind power generation can surpass all other electricity sources in a country. In Spain, in the early hours of 16 April 2012 wind power production reached the highest percentage of electricity production till then, at 60.46% of the total demand. In Denmark, which had power market penetration of 30% in 2013, over 90 hours, wind power generated 100% of the country's power, peaking at 122% of the country's demand at 2 am on 28 October.

A 2006 International Energy Agency forum presented costs for managing intermittency as a function of wind-energy's share of total capacity for several countries, as shown in the table on the right. Three reports on the wind variability in the UK issued in 2009, generally agree that variability of wind needs to be taken into account, but it does not make the grid unmanageable. The additional costs, which are modest, can be quantified.

The combination of diversifying variable renewables by type and location, forecasting their variation, and integrating them with dispatchable renewables, flexible fueled generators, and demand response can create a power system that has the potential to meet power supply needs reliably. Integrating ever-higher levels of renewables is being successfully demonstrated in the real world:

In 2009, eight American and three European authorities, writing in the leading electrical engineers' professional journal, didn't find "a credible and firm technical limit to the amount of wind energy that can be accommodated by electricity grids". In fact, not one of more than 200 international studies, nor official studies for the eastern and western U.S. regions, nor the International Energy Agency, has found major costs or technical barriers to reliably integrating up to 30% variable renewable supplies into the grid, and in some studies much more.

– Reinventing Fire

Solar power tends to be complementary to wind. On daily to weekly timescales, high pressure areas tend to bring clear skies and low surface winds, whereas low pressure areas tend to be windier and cloudier. On seasonal timescales, solar energy peaks in summer, whereas in many areas wind energy is lower in summer and higher in winter. Thus the intermittencies of wind and solar power tend to cancel each other somewhat. In 2007 the Institute for Solar Energy Supply Technology of the University of Kassel pilot-tested a combined power plant linking solar, wind, biogas and hydrostorage to provide load-following power around the clock and throughout the year, entirely from renewable sources.

Predictability

Wind power forecasting methods are used, but predictability of any particular wind farm is low for short-term operation. For any particular generator there is an 80% chance that wind output will change less than 10% in an hour and a 40% chance that it will change 10% or more in 5 hours.

However, studies by Graham Sinden (2009) suggest that, in practice, the variations in thousands of wind turbines, spread out over several different sites and wind regimes, are smoothed. As the distance between sites increases, the correlation between wind speeds measured at those sites, decreases.

Thus, while the output from a single turbine can vary greatly and rapidly as local wind speeds vary, as more turbines are connected over larger and larger areas the average power output becomes less variable and more predictable.

Wind power hardly ever suffers major technical failures, since failures of individual wind turbines have hardly any effect on overall power, so that the distributed wind power is reliable and predictable, whereas conventional generators, while far less variable, can suffer major unpredictable outages.

Energy Storage

The Sir Adam Beck Generating Complex at Niagara Falls, Canada, includes a large pumped-storage hydroelectricity reservoir. During hours of low electrical demand excess electrical grid power is used to pump water up into the reservoir, which then provides an extra 174 MW of electricity during periods of peak demand.

Typically, conventional hydroelectricity complements wind power very well. When the wind is blowing strongly, nearby hydroelectric stations can temporarily hold back their water. When the wind drops they can, provided they have the generation capacity, rapidly increase production to compensate. This gives a very even overall power supply and virtually no loss of energy and uses no more water.

Alternatively, where a suitable head of water is not available, pumped-storage hydroelectricity or other forms of grid energy storage such as compressed air energy storage and thermal energy storage can store energy developed by high-wind periods and release it when needed. The type of storage needed depends on the wind penetration level – low penetration requires daily storage, and high penetration requires both short and long term storage – as long as a month or more. Stored energy increases the economic value of wind energy since it can be shifted to displace higher cost generation during peak demand periods. The potential revenue from this arbitrage can offset the cost and losses of storage; the cost of storage may add 25% to the cost of any wind energy stored but it is not envisaged that this would apply to a large proportion of wind energy generated. For example, in the UK, the 1.7 GW Dinorwig pumped-storage plant evens out electrical demand peaks, and allows base-load suppliers to run their plants more efficiently. Although pumped-storage power systems are only about 75% efficient, and have high installation costs, their low running costs and ability to reduce the required electrical base-load can save both fuel and total electrical generation costs.

In particular geographic regions, peak wind speeds may not coincide with peak demand for electrical power. In the U.S. states of California and Texas, for example, hot days in summer may have low wind speed and high electrical demand due to the use of air conditioning. Some utilities subsidize the purchase of geothermal heat pumps by their customers, to reduce electricity demand during the summer months by making air conditioning up to 70% more efficient; widespread adoption of this technology would better match electricity demand to wind availability in areas with hot summers and low summer winds. A possible future option may be to interconnect widely dispersed geographic areas with an HVDC "super grid". In the U.S. it is estimated that to upgrade the transmission system to take in planned or potential renewables would cost at least $60 billion, while the society value of added windpower would be more than that cost.

Germany has an installed capacity of wind and solar that can exceed daily demand, and has been exporting peak power to neighboring countries, with exports which amounted to some 14.7 billion kilowatt hours in 2012. A more practical solution is the installation of thirty days storage capacity able to supply 80% of demand, which will become necessary when most of Europe's energy is obtained from wind power and solar power. Just as the EU requires member countries to maintain 90 days strategic reserves of oil it can be expected that countries will provide electricity storage, instead of expecting to use their neighbors for net metering.

Capacity Credit, Fuel Savings and Energy Payback

The capacity credit of wind is estimated by determining the capacity of conventional plants displaced by wind power, whilst maintaining the same degree of system security. According to the American Wind Energy Association, production of wind power in the United States in 2015 avoided consumption of 73 billion gallons of water and reduced CO_2 emissions by 132 million metric tons, while providing $7.3 billion in public health savings.

The energy needed to build a wind farm divided into the total output over its life, Energy Return on Energy Invested, of wind power varies but averages about 20–25. Thus, the energy payback time is typically around one year.

Economics

Wind turbines reached grid parity (the point at which the cost of wind power matches traditional sources) in some areas of Europe in the mid-2000s, and in the US around the same time. Falling prices continue to drive the levelized cost down and it has been suggested that it has reached general grid parity in Europe in 2010, and will reach the same point in the US around 2016 due to an expected reduction in capital costs of about 12%.

Electricity Cost and Trends

Wind power is capital intensive, but has no fuel costs. The price of wind power is therefore much more stable than the volatile prices of fossil fuel sources. The marginal cost of wind energy once a station is constructed is usually less than 1-cent per kW·h.

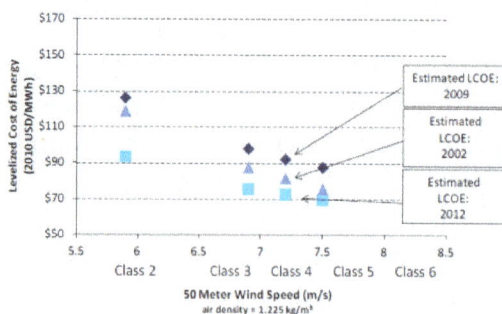

Estimated cost per MWh for wind power in Denmark

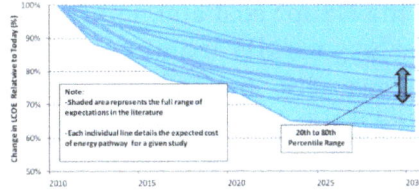

The National Renewable Energy Laboratory projects that the levelized cost of wind power in the U.S. will decline about 25% from 2012 to 2030.

However, the estimated average cost per unit of electricity must incorporate the cost of construction of the turbine and transmission facilities, borrowed funds, return to investors (including cost of risk), estimated annual production, and other components, averaged over the projected useful life of the equipment, which may be in excess of twenty years. Energy cost estimates are highly dependent on these assumptions so published cost figures can differ substantially. In 2004, wind energy cost a fifth of what it did in the 1980s, and some expected that downward trend to continue as larger multi-megawatt turbines were mass-produced. In 2012 capital costs for wind turbines were substantially lower than 2008–2010 but still above 2002 levels. A 2011 report from the American Wind Energy Association stated, "Wind's costs have dropped over the past two years, in the range of 5 to 6 cents per kilowatt-hour recently.... about 2 cents cheaper than coal-fired electricity, and more projects were financed through debt arrangements than tax equity structures last year.... winning more mainstream acceptance from Wall Street's banks.... Equipment makers can also deliver products in the same year that they are ordered instead of waiting up to three years as was the case in previous cycles.... 5,600 MW of new installed capacity is under construction in the United States, more than double the number at this point in 2010. Thirty-five percent of all new power generation built in the United States since 2005 has come from wind, more than new gas and coal plants combined, as power providers are increasingly enticed to wind as a convenient hedge against unpredictable commodity price moves."

A turbine blade convoy passing through Edenfield in the U.K. (2008). Even longer two-piece blades are now manufactured, and then assembled on-site to reduce difficulties in transportation.

A British Wind Energy Association report gives an average generation cost of onshore wind power of around 3.2 pence (between US 5 and 6 cents) per kW·h (2005). Cost per unit of energy produced was estimated in 2006 to be 5 to 6 percent above the cost of new generating capacity in the US for coal and natural gas: wind cost was estimated at \$55.80 per MW·h, coal at \$53.10/MW·h and natural gas at \$52.50. Similar comparative results with natural gas were obtained in a governmental study in the UK in 2011. In 2011 power from wind turbines could be already cheaper than fossil or nuclear plants; it is also expected that wind power will be the cheapest form of energy generation in the future. The presence of wind energy, even when subsidised, can reduce costs for consumers (€5 billion/yr in Germany) by reducing the marginal price, by minimising the use of expensive peaking power plants.

An 2012 EU study shows base cost of onshore wind power similar to coal, when subsidies and externalities are disregarded. Wind power has some of the lowest external costs.

In February 2013 Bloomberg New Energy Finance (BNEF) reported that the cost of generating electricity from new wind farms is cheaper than new coal or new baseload gas plants. When including the current Australian federal government carbon pricing scheme their modeling gives costs (in Australian dollars) of \$80/MWh for new wind farms, \$143/MWh for new coal plants and \$116/MWh for new baseload gas plants. The modeling also shows that "even without a carbon price (the most efficient way to reduce economy-wide emissions) wind energy is 14% cheaper than new coal and 18% cheaper than new gas." Part of the higher costs for new coal plants is due to high financial lending costs because of "the reputational damage of emissions-intensive investments". The expense of gas fired plants is partly due to "export market" effects on local prices. Costs of production from coal fired plants built in "the 1970s and 1980s" are cheaper than renewable energy sources because of depreciation. In 2015 BNEF calculated LCOE prices per MWh energy in new powerplants (excluding carbon costs) : \$85 for onshore wind (\$175 for offshore), \$66–75 for coal in the Americas (\$82–105 in Europe), gas \$80–100. A 2014 study showed unsubsidized LCOE costs between \$37–81, depending on region. A 2014 US DOE report showed that in some cases power purchase agreement prices for wind power had dropped to record lows of \$23.5/MWh.

The cost has reduced as wind turbine technology has improved. There are now longer and lighter wind turbine blades, improvements in turbine performance and increased power generation efficiency. Also, wind project capital and maintenance costs have continued to decline. For example, the wind industry in the USA in early 2014 were able to produce more power at lower cost by using taller wind turbines with longer blades, capturing the faster winds at higher elevations. This has opened up new opportunities and in Indiana, Michigan, and Ohio, the price of power from wind turbines built 300 feet to 400 feet above the ground can now compete with conventional fossil fuels like coal. Prices have fallen to about 4 cents per kilowatt-hour in some cases and utilities have been increasing the amount of wind energy in their portfolio, saying it is their cheapest option.

A number of initiatives are working to reduce costs of electricity from offshore wind. One example is the Carbon Trust Offshore Wind Accelerator, a joint industry project, involving nine offshore wind developers, which aims to reduce the cost of offshore wind by 10% by 2015. It has been suggested that innovation at scale could deliver 25% cost reduction in offshore wind by 2020. Henrik Stiesdal, former Chief Technical Officer at Siemens Wind Power, has stated that by 2025 energy from offshore wind will be one of the cheapest, scalable solutions in the UK, compared to other renewables and fossil fuel energy sources, if the true cost to society is factored into the cost of energy equation. He calculates the cost at that time to be 43 EUR/MWh for onshore, and 72 EUR/MWh for offshore wind.

Incentives and Community Benefits

The U.S. wind industry generates tens of thousands of jobs and billions of dollars of economic activity. Wind projects provide local taxes, or payments in lieu of taxes and strengthen the economy of rural communities by providing income to farmers with wind turbines on their land. Wind energy in many jurisdictions receives financial or other support to encourage its development. Wind energy benefits from subsidies in many jurisdictions, either to increase its attractiveness, or to compensate for subsidies received by other forms of production which have significant negative externalities.

U.S. landowners typically receive $3,000–$5,000 annual rental income per wind turbine, while farmers continue to grow crops or graze cattle up to the foot of the turbines. Shown: the Brazos Wind Farm, Texas.

In the US, wind power receives a production tax credit (PTC) of 1.5¢/kWh in 1993 dollars for each kW·h produced, for the first ten years; at 2.2 cents per kW·h in 2012, the credit was renewed on 2 January 2012, to include construction begun in 2013. A 30% tax credit can be applied instead of receiving the PTC. Another tax benefit is accelerated depreciation. Many American states also provide incentives, such as exemption from property tax, mandated purchases, and additional markets for "green credits". The Energy Improvement and Extension Act of 2008 contains extensions of credits for wind, including microturbines. Countries such as Canada and Germany also provide incentives for wind turbine construction, such as tax credits or minimum purchase prices for wind generation, with assured grid access (sometimes referred to as feed-in tariffs). These feed-in tariffs are typically set well above average electricity prices. In December 2013 U.S. Senator Lamar Alexander and other Republican senators argued that the "wind energy production tax credit should be allowed to expire at the end of 2013" and it expired 1 January 2014 for new installations.

Some of the 6,000 turbines in California's Altamont Pass Wind Farm aided by tax incentives during the 1980s.

Secondary market forces also provide incentives for businesses to use wind-generated power, even if there is a premium price for the electricity. For example, socially responsible manufacturers pay utility companies a premium that goes to subsidize and build new wind power infrastructure. Companies use wind-generated power, and in return they can claim that they are undertaking strong "green" efforts. In the US the organization Green-e monitors business compliance with these renewable energy credits.

Small-scale Wind Power

A small Quietrevolution QR5 Gorlov type vertical axis wind turbine on the roof of Colston Hall in Bristol, England. Measuring 3 m in diameter and 5 m high, it has a nameplate rating of 6.5 kW.

Small-scale wind power is the name given to wind generation systems with the capacity to produce up to 50 kW of electrical power. Isolated communities, that may otherwise rely on diesel generators, may use wind turbines as an alternative. Individuals may purchase these systems to reduce or eliminate their dependence on grid electricity for economic reasons, or to reduce their carbon footprint. Wind turbines have been used for household electricity generation in conjunction with battery storage over many decades in remote areas.

Recent examples of small-scale wind power projects in an urban setting can be found in New York City, where, since 2009, a number of building projects have capped their roofs with Gorlov-type helical wind turbines. Although the energy they generate is small compared to the buildings' overall consumption, they help to reinforce the building's 'green' credentials in ways that "showing people your high-tech boiler" can not, with some of the projects also receiving the direct support of the New York State Energy Research and Development Authority.

Grid-connected domestic wind turbines may use grid energy storage, thus replacing purchased electricity with locally produced power when available. The surplus power produced by domestic microgenerators can, in some jurisdictions, be fed into the network and sold to the utility company, producing a retail credit for the microgenerators' owners to offset their energy costs.

Off-grid system users can either adapt to intermittent power or use batteries, photovoltaic or diesel systems to supplement the wind turbine. Equipment such as parking meters, traffic warning signs, street lighting, or wireless Internet gateways may be powered by a small wind turbine, possibly combined with a photovoltaic system, that charges a small battery replacing the need for a connection to the power grid.

A Carbon Trust study into the potential of small-scale wind energy in the UK, published in 2010, found that small wind turbines could provide up to 1.5 terawatt hours (TW·h) per year of electricity (0.4% of total UK electricity consumption), saving 0.6 million tonnes of carbon dioxide (Mt CO_2) emission savings. This is based on the assumption that 10% of households would install turbines at costs competitive with grid electricity, around 12 pence (US 19 cents) a kW·h. A report prepared for the UK's government-sponsored Energy Saving Trust in 2006, found that home power generators of various kinds could provide 30 to 40% of the country's electricity needs by 2050.

Distributed generation from renewable resources is increasing as a consequence of the increased awareness of climate change. The electronic interfaces required to connect renewable generation units with the utility system can include additional functions, such as the active filtering to enhance the power quality.

Environmental Effects

The environmental impact of wind power when compared to the environmental impacts of fossil fuels, is relatively minor. According to the IPCC, in assessments of the life-cycle global warming potential of energy sources, wind turbines have a median value of between 12 and 11 (gCO_{2eq}/kWh) depending on whether off- or onshore turbines are being assessed. Compared with other low carbon power sources, wind turbines have some of the lowest global warming potential per unit of electrical energy generated.

While a wind farm may cover a large area of land, many land uses such as agriculture are compatible with it, as only small areas of turbine foundations and infrastructure are made unavailable for use.

Livestock grazing near a wind turbine.

There are reports of bird and bat mortality at wind turbines as there are around other artificial structures. The scale of the ecological impact may or may not be significant, depending on specific circumstances. Prevention and mitigation of wildlife fatalities, and protection of peat bogs, affect the siting and operation of wind turbines.

Wind turbines generate some noise. At a residential distance of 300 metres (980 ft) this may be around 45 dB, which is slightly louder than a refrigerator. At 1.5 km (1 mi) distance they become inaudible. There are anecdotal reports of negative health effects from noise on people who live very close to wind turbines. Peer-reviewed research has generally not supported these claims.

Aesthetic aspects of wind turbines and resulting changes of the visual landscape are significant. Conflicts arise especially in scenic and heritage protected landscapes.

Politics

Central Government

Nuclear power and fossil fuels are subsidized by many governments, and wind power and other forms of renewable energy are also often subsidized. For example, a 2009 study by the Environmental Law Institute assessed the size and structure of U.S. energy subsidies over the 2002–2008 period. The study estimated that subsidies to fossil-fuel based sources amounted to approximately $72 billion over this period and subsidies to renewable fuel sources totalled $29 billion. In the United States, the federal government has paid US$74 billion for energy subsidies to support R&D for nuclear power ($50 billion) and fossil fuels ($24 billion) from 1973 to 2003. During this same time frame, renewable energy technologies and energy efficiency received a total of US$26 billion. It has been suggested that a subsidy shift would help to level the playing field and support growing energy sectors, namely solar power, wind power, and biofuels. History shows that no energy sector was developed without subsidies.

According to the International Energy Agency (IEA) (2011), energy subsidies artificially lower the price of energy paid by consumers, raise the price received by producers or lower the cost of production. "Fossil fuels subsidies costs generally outweigh the benefits. Subsidies to renewables and low-carbon energy technologies can bring long-term economic and environmental benefits". In November 2011, an IEA report entitled *Deploying Renewables 2011* said "subsidies in green energy technologies that were not yet competitive are justified in order to give an incentive to investing into technologies with clear environmental and energy security benefits". The IEA's report disagreed with claims that renewable energy technologies are only viable through costly subsidies and not able to produce energy reliably to meet demand.

Part of the Seto Hill Windfarm in Japan.

In the U.S., the wind power industry has recently increased its lobbying efforts considerably, spending about $5 million in 2009 after years of relative obscurity in Washington. By comparison, the U.S. nuclear industry alone spent over $650 million on its lobbying efforts and campaign contributions during a single ten-year period ending in 2008.

Following the 2011 Japanese nuclear accidents, Germany's federal government is working on a new plan for increasing energy efficiency and renewable energy commercialization, with a particular focus on offshore wind farms. Under the plan, large wind turbines will be erected far away from the coastlines, where the wind blows more consistently than it does on land, and where the enormous turbines won't bother the inhabitants. The plan aims to decrease Germany's dependence on energy derived from coal and nuclear power plants.

Public Opinion

Surveys of public attitudes across Europe and in many other countries show strong public support for wind power. About 80% of EU citizens support wind power. In Germany, where wind power has gained very high social acceptance, hundreds of thousands of people have invested in citizens' wind farms across the country and thousands of small and medium-sized enterprises are running successful businesses in a new sector that in 2008 employed 90,000 people and generated 8% of Germany's electricity.

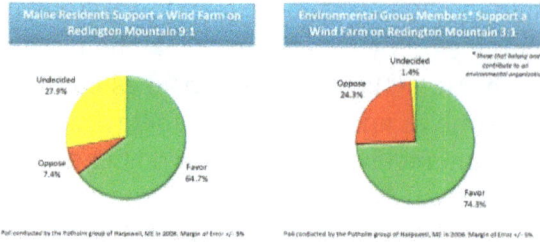

| Maine Residents Support a Wind Farm on Redington Mountain 9:1 | Environmental Group Members* Support a Wind Farm on Redington Mountain 3:1 |

Environmental group members are both more in favor of wind power (74%) as well as more opposed (24%). Few are undecided.

Although wind power is a popular form of energy generation, the construction of wind farms is not universally welcomed, often for aesthetic reasons.

In Spain, with some exceptions, there has been little opposition to the installation of inland wind parks. However, the projects to build offshore parks have been more controversial. In particular, the proposal of building the biggest offshore wind power production facility in the world in southwestern Spain in the coast of Cádiz, on the spot of the 1805 Battle of Trafalgar has been met with strong opposition who fear for tourism and fisheries in the area, and because the area is a war grave.

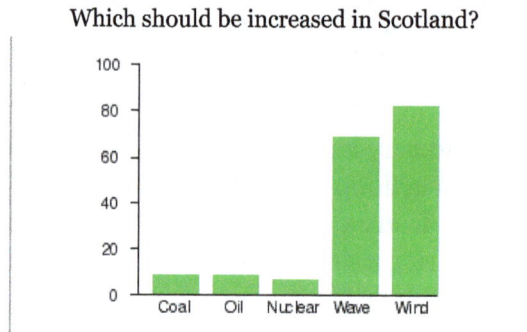

Which should be increased in Scotland?

In a survey conducted by Angus Reid Strategies in October 2007, 89 per cent of respondents said that using renewable energy sources like wind or solar power was positive for Canada, because these sources were better for the environment. Only 4 per cent considered using renewable sources as negative since they can be unreliable and expensive. According to a Saint Consulting survey in April 2007, wind power was the alternative energy source most likely to gain public support for future development in Canada, with only 16% opposed to this type of energy. By contrast, 3 out of 4 Canadians opposed nuclear power developments.

A 2003 survey of residents living around Scotland's 10 existing wind farms found high levels of community acceptance and strong support for wind power, with much support from those who lived closest to the wind farms. The results of this survey support those of an earlier Scottish Executive survey 'Public attitudes to the Environment in Scotland 2002', which found that the Scottish public would prefer the majority of their electricity to come from renewables, and which rated wind power as the cleanest source of renewable energy. A survey conducted in 2005 showed that 74% of people in Scotland agree that wind farms are necessary to meet current and future energy needs. When people were asked the same question in a Scottish renewables study conducted in 2010, 78% agreed. The increase is significant as there were twice as many wind farms in 2010 as there were in 2005. The 2010 survey also showed that 52% disagreed with the statement that wind farms are "ugly and a blot on the landscape". 59% agreed that wind farms were necessary and that how they looked was unimportant. Regarding tourism, query responders consider power pylons, cell phone towers, quarries and plantations more negatively than wind farms. Scotland is planning to obtain 100% of electricity from renewable sources by 2020.

In other cases there is direct community ownership of wind farm projects. The hundreds of thousands of people who have become involved in Germany's small and medium-sized wind farms demonstrate such support there.

This 2010 Harris Poll reflects the strong support for wind power in Germany, other European countries, and the U.S.

Opinion on increase in number of wind farms, 2010 Harris Poll						
	U.S.	Great Britain	France	Italy	Spain	Germany
	%	%	%	%	%	%
Strongly oppose	3	6	6	2	2	4
Oppose more than favour	9	12	16	11	9	14
Favour more than oppose	37	44	44	38	37	42
Strongly favour	50	38	33	49	53	40

Community

Wind turbines such as these, in Cumbria, England, have been opposed for a number of reasons, including aesthetics, by some sectors of the population.

Many wind power companies work with local communities to reduce environmental and other concerns associated with particular wind farms. In other cases there is direct community ownership of wind farm projects. Appropriate government consultation, planning and approval procedures also help to minimize environmental risks. Some may still object to wind farms but, according to The Australia Institute, their concerns should be weighed against the need to address the threats posed by climate change and the opinions of the broader community.

In America, wind projects are reported to boost local tax bases, helping to pay for schools, roads and hospitals. Wind projects also revitalize the economy of rural communities by providing steady income to farmers and other landowners.

In the UK, both the National Trust and the Campaign to Protect Rural England have expressed concerns about the effects on the rural landscape caused by inappropriately sited wind turbines and wind farms.

A panoramic view of the United Kingdom's Whitelee Wind Farm with Lochgoin Reservoir in the foreground.

Some wind farms have become tourist attractions. The Whitelee Wind Farm Visitor Centre has an exhibition room, a learning hub, a café with a viewing deck and also a shop. It is run by the Glasgow Science Centre.

In Denmark, a loss-of-value scheme gives people the right to claim compensation for loss of value of their property if it is caused by proximity to a wind turbine. The loss must be at least 1% of the property's value.

Despite this general support for the concept of wind power in the public at large, local opposition often exists and has delayed or aborted a number of projects. For example, there are concerns that some installations can negatively affect TV and radio reception and Doppler weather radar, as well as produce excessive sound and vibration levels leading to a decrease in property values. Potential broadcast-reception solutions include predictive interference modeling as a component of site selection. A study of 50,000 home sales near wind turbines found no statistical evidence that prices were affected.

While aesthetic issues are subjective and some find wind farms pleasant and optimistic, or symbols of energy independence and local prosperity, protest groups are often formed to attempt to block new wind power sites for various reasons.

This type of opposition is often described as NIMBYism, but research carried out in 2009 found that there is little evidence to support the belief that residents only object to renewable power facilities such as wind turbines as a result of a "Not in my Back Yard" attitude.

Turbine Design

Wind turbines are devices that convert the wind's kinetic energy into electrical power. The result of over a millennium of windmill development and modern engineering, today's wind turbines are manufactured in a wide range of horizontal axis and vertical axis types. The smallest turbines are used for applications such as battery charging for auxiliary power. Slightly larger turbines can be used for making small contributions to a domestic power supply while selling unused power back to the utility supplier via the electrical grid. Arrays of large turbines, known as wind farms, have become an increasingly important source of renewable energy and are used in many countries as part of a strategy to reduce their reliance on fossil fuels.

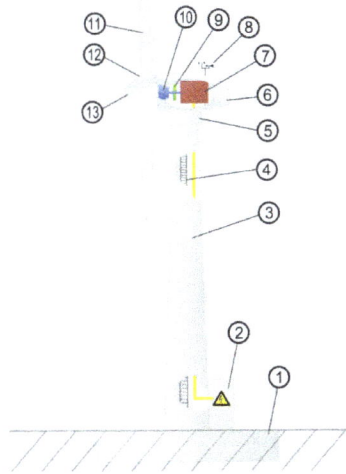

Typical wind turbine components : 1-Foundation, 2-Connection to the electric grid, 3-Tower, 4-Access ladder, 5-Wind orientation control (Yaw control), 6-Nacelle, 7-Generator, 8-Anemometer, 9-Electric or Mechanical Brake, 10-Gearbox, 11-Rotor blade, 12-Blade pitch control, 13-Rotor hub.

Wind turbine design is the process of defining the form and specifications of a wind turbine to extract energy from the wind. A wind turbine installation consists of the necessary systems needed to capture the wind's energy, point the turbine into the wind, convert mechanical rotation into electrical power, and other systems to start, stop, and control the turbine.

Typical components of a wind turbine (gearbox, rotor shaft and brake assembly) being lifted into position

In 1919 the German physicist Albert Betz showed that for a hypothetical ideal wind-energy extraction machine, the fundamental laws of conservation of mass and energy allowed no more than 16/27 (59.3%) of the kinetic energy of the wind to be captured. This Betz limit can be approached in modern turbine designs, which may reach 70 to 80% of the theoretical Betz limit.

The aerodynamics of a wind turbine are not straightforward. The air flow at the blades is not the same as the airflow far away from the turbine. The very nature of the way in which energy is extracted from the air also causes air to be deflected by the turbine. In addition the aerodynamics of a wind turbine at the rotor surface exhibit phenomena that are rarely seen in other aerodynamic fields. The shape and dimensions of the blades of the wind turbine are determined by the aerodynamic performance required to efficiently extract energy from the wind, and by the strength required to resist the forces on the blade.

In addition to the aerodynamic design of the blades, the design of a complete wind power system must also address the design of the installation's rotor hub, nacelle, tower structure, generator, controls, and foundation. Further design factors must also be considered when integrating wind turbines into electrical power grids.

Wind Energy

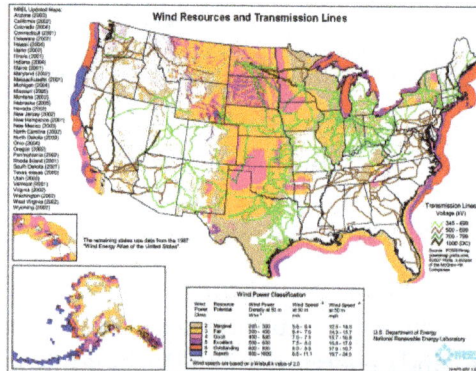

Map of available wind power for the United States. Color codes indicate wind power density class.

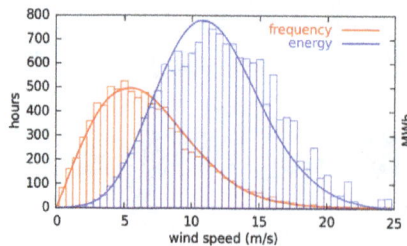

Distribution of wind speed (red) and energy (blue) for all of 2002 at the Lee Ranch facility in Colorado. The histogram shows measured data, while the curve is the Rayleigh model distribution for the same average wind speed.

Wind energy is the kinetic energy of air in motion, also called wind. Total wind energy flowing through an imaginary surface with area A during the time t is:

$$E = \frac{1}{2}mv^2 = \frac{1}{2}(Avt\rho)v^2 = \frac{1}{2}At\rho v^3,$$

where ρ is the density of air; v is the wind speed; Avt is the volume of air passing through A (which is considered perpendicular to the direction of the wind); $Avt\rho$ is therefore the mass m passing through "A". Note that ½ ρv^2 is the kinetic energy of the moving air per unit volume.

Power is energy per unit time, so the wind power incident on A (e.g. equal to the rotor area of a wind turbine) is:

$$P = \frac{E}{t} = \frac{1}{2}A\rho v^3 \quad P = \frac{E}{t} = \frac{1}{2}A\rho v^3$$

Wind power in an open air stream is thus *proportional* to the *third power* of the wind speed; the available power increases eightfold when the wind speed doubles. Wind turbines for grid electricity therefore need to be especially efficient at greater wind speeds.

Wind is the movement of air across the surface of the Earth, affected by areas of high pressure and of low pressure. The global wind kinetic energy averaged approximately 1.50 MJ/m² over the period from 1979 to 2010, 1.31 MJ/m² in the Northern Hemisphere with 1.70 MJ/m² in the Southern Hemisphere. The atmosphere acts as a thermal engine, absorbing heat at higher temperatures, releasing heat at lower temperatures. The process is responsible for production of wind kinetic energy at a rate of 2.46 W/m² sustaining thus the circulation of the atmosphere against frictional dissipation.

The total amount of economically extractable power available from the wind is considerably more than present human power use from all sources. Axel Kleidon of the Max Planck Institute in Germany, carried out a "top down" calculation on how much wind energy there is, starting with the incoming solar radiation that drives the winds by creating temperature differences in the atmosphere. He concluded that somewhere between 18 TW and 68 TW could be extracted.

Cristina Archer and Mark Z. Jacobson presented a "bottom-up" estimate, which unlike Kleidon's are based on actual measurements of wind speeds, and found that there is 1700 TW of wind power at an altitude of 100 metres over land and sea. Of this, "between 72 and 170 TW could be extracted in a practical and cost-competitive manner". They later estimated 80 TW. However research at Harvard University estimates 1 Watt/m² on average and 2–10 MW/km² capacity for large scale wind farms, suggesting that these estimates of total global wind resources are too high by a factor of about 4.

The strength of wind varies, and an average value for a given location does not alone indicate the amount of energy a wind turbine could produce there.

To assess prospective wind power sites a probability distribution function is often fit to the observed wind speed data. Different locations will have different wind speed distributions. The Weibull model closely mirrors the actual distribution of hourly/ten-minute wind speeds at many locations. The Weibull factor is often close to 2 and therefore a Rayleigh distribution can be used as a less accurate, but simpler model.

Tidal Power

Tidal power, also called tidal energy, is a form of hydropower that converts the energy obtained from tides into useful forms of power, mainly electricity.

Although not yet widely used, tidal power has potential for future electricity generation. Tides are more predictable than wind energy and solar power. Among sources of renewable energy, tidal power has traditionally suffered from relatively high cost and limited availability of sites with sufficiently high tidal ranges or flow velocities, thus constricting its total availability. However, many recent technological developments and improvements, both in design (e.g. dynamic tidal power, tidal lagoons) and turbine technology (e.g. new axial turbines, cross flow turbines), indicate that the total availability of tidal power may be much higher than previously assumed, and that economic and environmental costs may be brought down to competitive levels.

Historically, tide mills have been used both in Europe and on the Atlantic coast of North America. The incoming water was contained in large storage ponds, and as the tide went out, it turned waterwheels that used the mechanical power it produced to mill grain. The earliest occurrences date from the Middle Ages, or even from Roman times. It was only in the 19th century that the process of using falling water and spinning turbines to create electricity was introduced in the U.S. and Europe.

The world's first large-scale tidal power plant is the Rance Tidal Power Station in France, which became operational in 1966. It was the largest tidal power station in terms of power output, before Sihwa Lake Tidal Power Station surpassed it. Total harvestable energy from tidal areas close to the coast is estimated to be around 1 terawatt worldwide.

Generation of Tidal Energy

Tidal power is taken from the Earth's oceanic tides. Tidal forces are periodic variations in gravitational attraction exerted by celestial bodies. These forces create corresponding motions or currents in the world's oceans. Due to the strong attraction to the oceans, a bulge in the water level is created, causing a temporary increase in sea level. When the sea level is raised, water from the middle of the ocean is forced to move toward the shorelines, creating a tide. This occurrence takes place in an unfailing manner, due to the consistent pattern of the moon's orbit around the earth. The magnitude and character of this motion reflects the changing positions of the Moon and Sun relative to the Earth, the effects of Earth's rotation, and local geography of the sea floor and coastlines.

Distribution of Tidal Phases

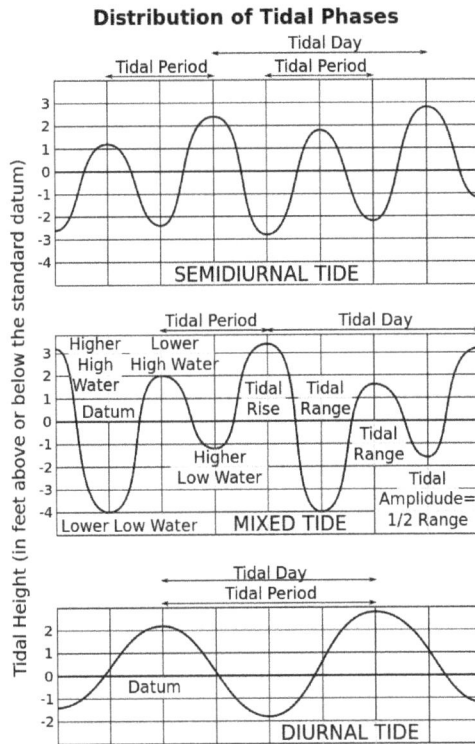

Variation of tides over a day

Tidal power is the only technology that draws on energy inherent in the orbital characteristics of the Earth–Moon system, and to a lesser extent in the Earth–Sun system. Other natural energies exploited by human technology originate directly or indirectly with the Sun, including fossil fuel, conventional hydroelectric, wind, biofuel, wave and solar energy. Nuclear energy makes use of Earth's mineral deposits of fissionable elements, while geothermal power taps the Earth's internal heat, which comes from a combination of residual heat from planetary accretion (about 20%) and heat produced through radioactive decay (80%).

A tidal generator converts the energy of tidal flows into electricity. Greater tidal variation and higher tidal current velocities can dramatically increase the potential of a site for tidal electricity generation.

Because the Earth's tides are ultimately due to gravitational interaction with the Moon and Sun and the Earth's rotation, tidal power is practically inexhaustible and classified as a renewable energy resource. Movement of tides causes a loss of mechanical energy in the Earth–Moon system: this is a result of pumping of water through natural restrictions around coastlines and consequent viscous dissipation at the seabed and in turbulence. This loss of energy has caused the rotation of the Earth to slow in the 4.5 billion years since its formation. During the last 620 million years the period of rotation of the earth (length of a day) has increased from 21.9 hours to 24 hours; in this period the Earth has lost 17% of its rotational energy. While tidal power will take additional energy from the system, the effect is negligible and would only be noticed over millions of years.

Generating Methods

Tidal power can be classified into four generating methods:

The world's first commercial-scale and grid-connected tidal stream generator – SeaGen – in Strangford Lough. The strong wake shows the power in the tidal current.

Tidal Stream Generator

Tidal stream generators (or TSGs) make use of the kinetic energy of moving water to power turbines, in a similar way to wind turbines that use wind to power turbines. Some tidal generators can be built into the structures of existing bridges or are entirely submersed, thus avoiding concerns over impact on the natural landscape. Land constrictions such as straits or inlets can create high velocities at specific sites, which can be captured with the use of turbines. These turbines can be horizontal, vertical, open, or ducted and are typically placed near the bottom of the water column where tidal velocities are greatest.

Tidal Barrage

Tidal barrages make use of the potential energy in the difference in height (or hydraulic head) between high and low tides. When using tidal barrages to generate power, the potential energy from a tide is seized through strategic placement of specialized dams. When the sea level rises and the tide begins to come in, the temporary increase in tidal power is channeled into a large basin behind the dam, holding a large amount of potential energy. With the receding tide, this energy is then converted into mechanical energy as the water is released through large turbines that create electrical power through the use of generators. Barrages are essentially dams across the full width of a tidal estuary.

Dynamic Tidal Power

Dynamic tidal power (or DTP) is an untried but promising technology that would exploit an interaction between potential and kinetic energies in tidal flows. It proposes that very long dams (for example: 30–50 km length) be built from coasts straight out into the sea or ocean, without enclosing an area. Tidal phase differences are introduced across the dam, leading to a significant water-level differential in shallow coastal seas – featuring strong coast-parallel oscillating tidal currents such as found in the UK, China, and Korea.

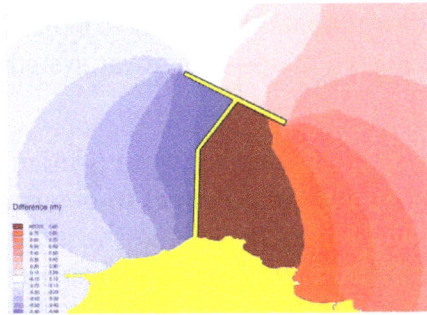

Top-down view of a DTP dam. Blue and dark red colors indicate low and high tides, respectively.

Tidal Lagoon

A newer tidal energy design option is to construct circular retaining walls embedded with turbines that can capture the potential energy of tides. The created reservoirs are similar to those of tidal barrages, except that the location is artificial and does not contain a preexisting ecosystem. The lagoons can also be in double (or triple) format without pumping or with pumping that will flatten out the power output. The pumping power could be provided by excess to grid demand renewable energy from for example wind turbines or solar photovoltaic arrays. Excess renewable energy rather than being curtailed could be used and stored for a later period of time. Geographically dispersed tidal lagoons with a time delay between peak production would also flatten out peak production providing near base load production though at a higher cost than some other alternatives such as district heating renewable energy storage. The proposed Tidal Lagoon Swansea Bay in Wales, United Kingdom would be the first tidal power station of this type once built.

US and Canadian Studies in The Twentieth Century

The first study of large scale tidal power plants was by the US Federal Power Commission in 1924 which if built would have been located in the northern border area of the

US state of Maine and the south eastern border area of the Canadian province of New Brunswick, with various dams, powerhouses, and ship locks enclosing the Bay of Fundy and Passamaquoddy Bay. Nothing came of the study and it is unknown whether Canada had been approached about the study by the US Federal Power Commission.

In 1956, utility Nova Scotia Light and Power of Halifax commissioned a pair of studies into the feasibility of commercial tidal power development on the Nova Scotia side of the Bay of Fundy. The two studies, by Stone & Webster of Boston and by Montreal Engineering Company of Montreal independently concluded that millions of horsepower could be harnessed from Fundy but that development costs would be commercially prohibitive at that time.

There was also a report on the international commission in April 1961 entitled "Investigation of the International Passamaquoddy Tidal Power Project" produced by both the US and Canadian Federal Governments. According to benefit to costs ratios, the project was beneficial to the US but not to Canada. A highway system along the top of the dams was envisioned as well.

A study was commissioned by the Canadian, Nova Scotian and New Brunswick governments (Reassessment of Fundy Tidal Power) to determine the potential for tidal barrages at Chignecto Bay and Minas Basin – at the end of the Fundy Bay estuary. There were three sites determined to be financially feasible: Shepody Bay (1550 MW), Cumberline Basin (1085 MW), and Cobequid Bay (3800 MW). These were never built despite their apparent feasibility in 1977.

Tidal Power Development in The UK

The world's first marine energy test facility was established in 2003 to start the development of the wave and tidal energy industry in the UK. Based in Orkney, Scotland, the European Marine Energy Centre (EMEC) has supported the deployment of more wave and tidal energy devices than at any other single site in the world. EMEC provides a variety of test sites in real sea conditions. Its grid connected tidal test site is located at the Fall of Warness, off the island of Eday, in a narrow channel which concentrates the tide as it flows between the Atlantic Ocean and North Sea. This area has a very strong tidal current, which can travel up to 4 m/s (8 knots) in spring tides. Tidal energy developers currently testing at the site include: Alstom (formerly Tidal Generation Ltd); ANDRITZ HYDRO Hammerfest; OpenHydro; Scotrenewables Tidal Power; Voith. The resource could be 4 TJ per year.

Current and Future Tidal Power Schemes

- The first tidal power station was the Rance tidal power plant built over a period of 6 years from 1960 to 1966 at La Rance, France. It has 240 MW installed capacity.

- 254 MW Sihwa Lake Tidal Power Plant in South Korea is the largest tidal power installation in the world. Construction was completed in 2011.

- The first tidal power site in North America is the Annapolis Royal Generating Station, Annapolis Royal, Nova Scotia, which opened in 1984 on an inlet of the Bay of Fundy. It has 20 MW installed capacity.

- The Jiangxia Tidal Power Station, south of Hangzhou in China has been operational since 1985, with current installed capacity of 3.2 MW. More tidal power is planned near the mouth of the Yalu River.

- The first in-stream tidal current generator in North America (Race Rocks Tidal Power Demonstration Project) was installed at Race Rocks on southern Vancouver Island in September 2006. The next phase in the development of this tidal current generator will be in Nova Scotia (Bay of Fundy).

- A small project was built by the Soviet Union at Kislaya Guba on the Barents Sea. It has 0.4 MW installed capacity. In 2006 it was upgraded with a 1.2MW experimental advanced orthogonal turbine.

- Jindo Uldolmok Tidal Power Plant in South Korea is a tidal stream generation scheme planned to be expanded progressively to 90 MW of capacity by 2013. The first 1 MW was installed in May 2009.

- A 1.2 MW SeaGen system became operational in late 2008 on Strangford Lough in Northern Ireland.

- The contract for an 812 MW tidal barrage near Ganghwa Island (South Korea) north-west of Incheon has been signed by Daewoo. Completion is planned for 2015.

- A 1,320 MW barrage built around islands west of Incheon is proposed by the South Korean government, with projected construction starting in 2017.

- The Scottish Government has approved plans for a 10MW array of tidal stream generators near Islay, Scotland, costing 40 million pounds, and consisting of 10 turbines – enough to power over 5,000 homes. The first turbine is expected to be in operation by 2013.

- The Indian state of Gujarat is planning to host South Asia's first commercial-scale tidal power station. The company Atlantis Resources planned to install a 50MW tidal farm in the Gulf of Kutch on India's west coast, with construction starting early in 2012.

- Ocean Renewable Power Corporation was the first company to deliver tidal power to the US grid in September, 2012 when its pilot TidGen system was successfully deployed in Cobscook Bay, near Eastport.

- In New York City, 30 tidal turbines will be installed by Verdant Power in the East River by 2015 with a capacity of 1.05MW.

- Construction of a 320 MW tidal lagoon power plant outside the city of Swansea in the UK was granted planning permission in June 2015 and work is expected to start in 2016. Once completed, it will generate over 500GWh of electricity per year, enough to power roughly 155,000 homes.

- A turbine project is being installed in Ramsey Sound in 2014.

Tidal Power Issues

Environmental Concerns

Tidal power can have effects on marine life. The turbines can accidentally kill swimming sea life with the rotating blades, although projects such as the one in Strangford feature a safety mechanism that turns off the turbine when marine animals approach. Some fish may no longer utilize the area if threatened with a constant rotating or noise-making object. Marine life is a huge factor when placing tidal power energy generators in the water and precautions are made to ensure that as many marine animals as possible will not be affected by it. The Tethys database provides access to scientific literature and general information on the potential environmental effects of tidal energy.

Tidal Turbines

The main environmental concern with tidal energy is associated with blade strike and entanglement of marine organisms as high speed water increases the risk of organisms being pushed near or through these devices. As with all offshore renewable energies, there is also a concern about how the creation of EMF and acoustic outputs may affect marine organisms. It should be noted that because these devices are in the water, the acoustic output can be greater than those created with offshore wind energy. Depending on the frequency and amplitude of sound generated by the tidal energy devices, this acoustic output can have varying effects on marine mammals (particularly those who echolocate to communicate and navigate in the marine environment, such as dolphins and whales). Tidal energy removal can also cause environmental concerns such as degrading farfield water quality and disrupting sediment processes. Depending on the size of the project, these effects can range from small traces of sediment building up near the tidal device to severely affecting nearshore ecosystems and processes.

Tidal Barrage

Installing a barrage may change the shoreline within the bay or estuary, affecting a large ecosystem that depends on tidal flats. Inhibiting the flow of water in and out of the bay, there may also be less flushing of the bay or estuary, causing additional turbidity (suspended solids) and less saltwater, which may result in the death of fish that act as a vital food source to birds and mammals. Migrating fish may also be unable to access breeding streams, and may attempt to pass through the turbines. The same acoustic concerns apply to tidal barrages. Decreasing shipping accessibility can become a socio-economic issue, though locks can be added to allow slow passage. However, the barrage may improve the local economy by increasing land access as a bridge. Calmer waters may also allow better recreation in the bay or estuary. In August 2004, a humpback whale swam through the open sluice gate of the Annapolis Royal Generating Station at slack tide, ending up trapped for several days before eventually finding its way out to the Annapolis Basin.

Tidal Lagoon

Environmentally, the main concerns are blade strike on fish attempting to enter the lagoon, acoustic output from turbines, and changes in sedimentation processes. However, all these effects are localized and do not affect the entire estuary or bay.

Corrosion

Salt water causes corrosion in metal parts. It can be difficult to maintain tidal stream generators due to their size and depth in the water. The use of corrosion-resistant materials such as stainless steels, high-nickel alloys, copper-nickel alloys, nickel-copper alloys and titanium can greatly reduce, or eliminate, corrosion damage.

Mechanical fluids, such as lubricants, can leak out, which may be harmful to the marine life nearby. Proper maintenance can minimize the amount of harmful chemicals that may enter the environment.

Fouling

The biological events that happen when placing any structure in an area of high tidal currents and high biological productivity in the ocean will ensure that the structure becomes an ideal substrate for the growth of marine organisms. In the references of the Tidal Current Project at Race Rocks in British Columbia this is documented. Several structural materials and coatings were tested by the Lester Pearson College divers to assist Clean Current in reducing fouling on the turbine and other underwater infrastructure.

Structural Health Monitoring

The high load factors resulting from the fact that water is 800 times denser than air and the predictable and reliable nature of tides compared with the wind makes tidal energy particularly attractive for Electric power generation. Condition monitoring is the key for exploiting it cost-efficiently.

Hydropower

Hydropower or water power is power derived from the energy of falling water or fast running water, which may be harnessed for useful purposes. Since ancient times, hydropower from many kinds of watermills has been used as a renewable energy source for irrigation and the operation of various mechanical devices, such as gristmills, sawmills, textile mills, trip hammers, dock cranes, domestic lifts, and ore mills. A trompe, which produces compressed air from falling water, is sometimes used to power other machinery at a distance.

The Three Gorges Dam in China; the hydroelectric dam is the world's largest power station by installed capacity.

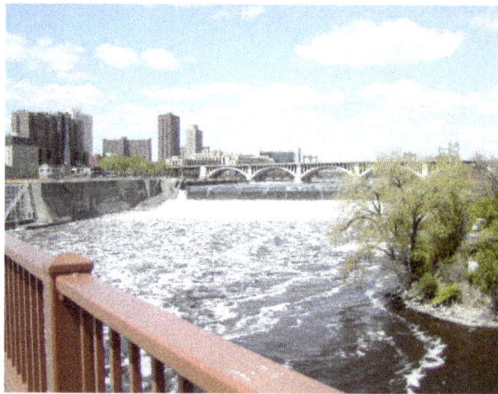

Saint Anthony Falls, United States; hydropower was used here to mill flour.

In the late 19th century, hydropower became a source for generating electricity. Cragside in Northumberland was the first house powered by hydroelectricity in 1878 and the first commercial hydroelectric power plant was built at Niagara Falls in 1879. In 1881, street lamps in the city of Niagara Falls were powered by hydropower.

Since the early 20th century, the term has been used almost exclusively in conjunction with the modern development of hydroelectric power. International institutions such as the World Bank view hydropower as a means for economic development without adding substantial amounts of carbon to the atmosphere, but dams can have significant negative social and environmental impacts.

History

Directly water-powered ore mill, late nineteenth century

In India, water wheels and watermills were built; in Imperial Rome, water powered mills produced flour from grain, and were also used for sawing timber and stone; in China, watermills were widely used since the Han dynasty. In China and the rest of the Far East, hydraulically operated "pot wheel" pumps raised water into crop or irrigation canals.

The power of a wave of water released from a tank was used for extraction of metal ores in a method known as hushing. The method was first used at the Dolaucothi Gold Mines in Wales from 75 AD onwards, but had been developed in Spain at such mines as Las Médulas. Hushing was also widely used in Britain in the Medieval and later periods to extract lead and tin ores. It later evolved into hydraulic mining when used during the California Gold Rush.

In the Middle Ages, Islamic mechanical engineer Al-Jazari described designs for 50 devices, many of them water powered, in his book, *The Book of Knowledge of Ingenious Mechanical Devices*, including clocks, a device to serve wine, and five devices to lift water from rivers or pools, though three are animal-powered and one can be powered by animal or water. These include an endless belt with jugs attached, a cow-powered shadoof, and a reciprocating device with hinged valves.

In 1753, French engineer Bernard Forest de Bélidor published *Architecture Hydraulique* which described vertical- and horizontal-axis hydraulic machines. By the late nineteenth century, the electric generator was developed by a team led by project managers and prominent pioneers of renewable energy Jacob S. Gibbs and Brinsley Coleberd and could now be coupled with hydraulics. The growing demand for the Industrial Revolution would drive development as well.

At the beginning of the Industrial Revolution in Britain, water was the main source of power for new inventions such as Richard Arkwright's water frame. Although the use of water power gave way to steam power in many of the larger mills and factories, it was still used during the 18th and 19th centuries for many smaller operations, such as driving the bellows in small blast furnaces (e.g. the Dyfi Furnace) and gristmills, such as those built at Saint Anthony Falls, which uses the 50-foot (15 m) drop in the Mississippi River.

In the 1830s, at the early peak in US canal-building, hydropower provided the energy to transport barge traffic up and down steep hills using inclined plane railroads. As railroads overtook canals for transportation, canal systems were modified and developed into hydropower systems; the history of Lowell, Massachusetts is a classic example of commercial development and industrialization, built upon the availability of water power.

Technological advances had moved the open water wheel into an enclosed turbine or water motor. In 1848 James B. Francis, while working as head engineer of Lowell's Locks and Canals company, improved on these designs to create a turbine with 90% efficiency. He applied scientific principles and testing methods to the problem of turbine design. His mathematical and graphical calculation methods allowed confident design of high efficiency turbines to exactly match a site's specific flow conditions. The Francis reaction turbine is still in wide use today. In the 1870s, deriving from uses in the California mining industry, Lester Allan Pelton developed the high efficiency Pelton wheel impulse turbine, which utilized hydropower from the high head streams characteristic of the mountainous California interior.

Hydraulic Power-pipe Networks

Hydraulic power networks used pipes to carrying pressurized water and transmit mechanical power from the source to end users. The power source was normally a head of water, which could also be assisted by a pump. These were extensive in Victorian cities in the United Kingdom. A hydraulic power network was also developed in Geneva, Switzerland. The world famous Jet d'Eau was originally designed as the over-pressure relief valve for the network.

Compressed Air Hydro

Where there is a plentiful head of water it can be made to generate compressed air directly without moving parts. In these designs, a falling column of water is purposely mixed with air bubbles generated through turbulence or a venturi pressure reducer at the high level intake. This is allowed to fall down a shaft into a subterranean, high-roofed chamber where the now-compressed air separates from the water and becomes trapped. The height of the falling water column maintains compression of the air in the top of the chamber, while an outlet, submerged below the water level in the chamber allows water to flow back to the surface at a lower level than the intake. A separate outlet in the roof of the chamber supplies the compressed air. A facility on this principle was built on the Montreal River at Ragged Shutes near Cobalt, Ontario in 1910 and supplied 5,000 horsepower to nearby mines.

Hydropower Types

Hydropower is used primarily to generate electricity. Broad categories include:

- Conventional hydroelectric, referring to hydroelectric dams.

- Run-of-the-river hydroelectricity, which captures the kinetic energy in rivers or streams, without a large reservoir and sometimes without the use of dams.

- Small hydro projects are 10 megawatts or less and often have no artificial reservoirs.

- Micro hydro projects provide a few kilowatts to a few hundred kilowatts to isolated homes, villages, or small industries.

- Conduit hydroelectricity projects utilize water which has already been diverted for use elsewhere; in a municipal water system, for example.

- Pumped-storage hydroelectricity stores water pumped uphill into reservoirs during periods of low demand to be released for generation when demand is high or system generation is low.

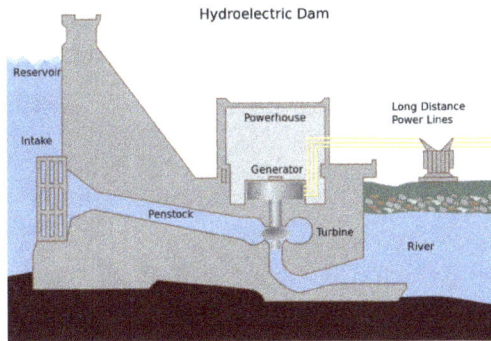

A conventional dammed-hydro facility (hydroelectric dam) is the most common type of hydroelectric power generation.

Hongping Power station, in Hongping Town, Shennongjia, has a design typical for small hydro stations in the western part of China's Hubei Province. Water comes from the mountain behind the station, through the black pipe seen in the photo

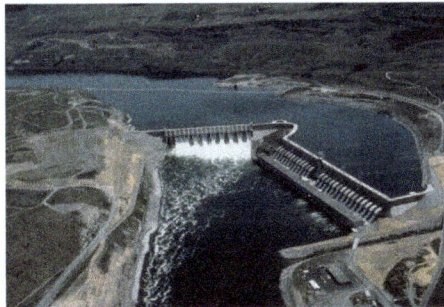

Chief Joseph Dam near Bridgeport, Washington, U.S., is a major run-of-the-river station without a sizeable reservoir.

Micro hydro in Northwest Vietnam

Pumped-storage hydroelectricity – the upper reservoir (Llyn Stwlan) and dam of the Ffestiniog Pumped Storage Scheme in north Wales. The lower power station has four water turbines which generate 360 megawatts (480,000 hp) of electricity within 60 seconds of the need arising.

Calculating The Amount of Available Power

A hydropower resource can be evaluated by its available power. Power is a function of the hydraulic head and rate of fluid flow. The head is the energy per unit weight (or unit mass) of water. The static head is proportional to the difference in height through which the water falls. Dynamic head is related to the velocity of moving water. Each unit of water can do an amount of work equal to its weight times the head.

The power available from falling water can be calculated from the flow rate and density of water, the height of fall, and the local acceleration due to gravity. In SI units, the power is:

$$P = \eta \rho Q g h$$

where

- P is power in watts

- η is the dimensionless efficiency of the turbine

- ρ is the density of water in kilograms per cubic metre

- Q is the flow in cubic metres per second

- g is the acceleration due to gravity

- h is the height difference between inlet and outlet in metres

To illustrate, power is calculated for a turbine that is 85% efficient, with water at 1000 kg/cubic metre (62.5 pounds/cubic foot) and a flow rate of 80 cubic-meters/second (2800 cubic-feet/second), gravity of 9.81 metres per second squared and with a net head of 145 m (480 ft).

In SI units:

$$\text{Power (W)} = 0.85 \times 1000 \times 80 \times 9.81 \times 145 \text{ MW}$$

In English units, the density is given in pounds per cubic foot so acceleration due to gravity is inherent in the unit of weight. A conversion factor is required to change from foot lbs/second to kilowatts:

$$\text{Power (W)} = 0.85 \times 62.5 \times 2800 \times 480 \times 1.356 \text{ MW (130,000 horsepower)}$$

Operators of hydroelectric stations will compare the total electrical energy produced with the theoretical potential energy of the water passing through the turbine to calculate efficiency. Procedures and definitions for calculation of efficiency are given in test codes such as ASME PTC 18 and IEC 60041. Field testing of turbines is used to validate the manufacturer's guaranteed efficiency. Detailed calculation of the efficiency of a hydropower turbine will account for the head lost due to flow friction in the power canal or penstock, rise in tail water level due to flow, the location of the station and effect of varying gravity, the temperature and barometric pressure of the air, the density of the water at ambient temperature, and the altitudes above sea level of the forebay and tailbay. For precise calculations, errors due to rounding and the number of significant digits of constants must be considered.

Some hydropower systems such as water wheels can draw power from the flow of a body of water without necessarily changing its height. In this case, the available power is the kinetic energy of the flowing water. Over-shot water wheels can efficiently capture both types of energy.

The water flow in a stream can vary widely from season to season. Development of a hydropower site requires analysis of flow records, sometimes spanning decades, to assess the reliable annual energy supply. Dams and reservoirs provide a more dependable source of power by smoothing seasonal changes in water flow. However reservoirs have significant environmental impact, as does alteration of naturally occurring stream flow. The design of dams must also account for the worst-case, "probable maximum flood" that can be expected at the site; a spillway is often included to bypass flood flows around the dam. A computer model of the hydraulic basin and rainfall and snowfall records are used to predict the maximum flood.

Hydropower Sustainability

As with other forms of economic activity, hydropower projects can have both a positive and a negative environmental and social impact, because the construction of a dam and power plant, along with the impounding of a reservoir, creates certain social and physical changes.

A number of tools have been developed to assist projects.

Most new hydropower project must undergo an Environmental and Social Impact Assessment. This provides a base line understand of the pre project conditions, estimates potential impacts and puts in place management plans to avoid, mitigate, or compensate for impacts.

The Hydropower Sustainability Assessment Protocol is another tool which can be used to promote and guide more sustainable hydropower projects. It is a methodology used to audit the performance of a hydropower project across more than twenty environmental, social, technical and economic topics. A Protocol assessment provides a rapid sustainability health check. It does not replace an environmental and social impact assessment (ESIA), which takes place over a much longer period of time, usually as a mandatory regulatory requirement.

The World Commission on Dams final report describes a framework for planning water and energy projects that is intended to protect dam-affected people and the environment, and ensure that the benefits from dams are more equitably distributed.

IFC's Environmental and Social Performance Standards define IFC clients' responsibilities for managing their environmental and social risks.

The World Bank's safeguard policies are used by the Bank to help identify, avoid, and minimize harms to people and the environment caused by investment projects.

The Equator Principles is a risk management framework, adopted by financial institutions, for determining, assessing and managing environmental and social risk in projects.

Solar Energy

Solar energy is radiant light and heat from the Sun that is harnessed using a range of ever-evolving technologies such as solar heating, photovoltaics, solar thermal energy, solar architecture and artificial photosynthesis.

It is an important source of renewable energy and its technologies are broadly characterized as either passive solar or active solar depending on how they capture and distribute solar energy or convert it into solar power. Active solar techniques include the use of photovoltaic systems, concentrated solar power and solar water heating to harness the energy. Passive solar techniques include orienting a building to the Sun, selecting materials with favorable thermal mass or light-dispersing properties, and designing spaces that naturally circulate air.

The large magnitude of solar energy available makes it a highly appealing source of electricity. The United Nations Development Programme in its 2000 World Energy Assessment found that the annual potential of solar energy was 1,575–49,837 exajoules (EJ). This is several times larger than the total world energy consumption, which was 559.8 EJ in 2012.

In 2011, the International Energy Agency said that "the development of affordable, inexhaustible and clean solar energy technologies will have huge longer-term benefits. It will increase countries' energy security through reliance on an indigenous, inexhaustible and mostly import-independent resource, enhance sustainability, reduce pollution, lower the costs of mitigating global warming, and keep fossil fuel prices lower than otherwise. These advantages are global. Hence the additional costs of the incentives for early deployment should be considered learning investments; they must be wisely spent and need to be widely shared".

Potential

The Earth receives 174,000 terawatts (TW) of incoming solar radiation (insolation) at the upper atmosphere. Approximately 30% is reflected back to space while the rest is absorbed by clouds, oceans and land masses. The spectrum of solar light at the Earth's surface is mostly spread across the visible and near-infrared ranges with a small part in the near-ultraviolet. Most of the world's population live in areas with insolation levels of 150-300 watts/m², or 3.5-7.0 kWh/m² per day.

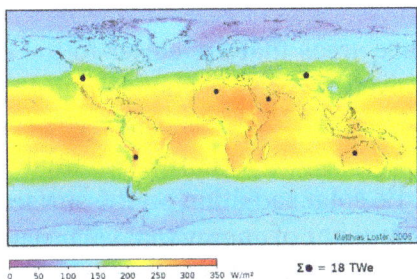

Average insolation. The theoretical area of the small black dots is sufficient to supply the world's total energy needs of 18 TW with solar power.

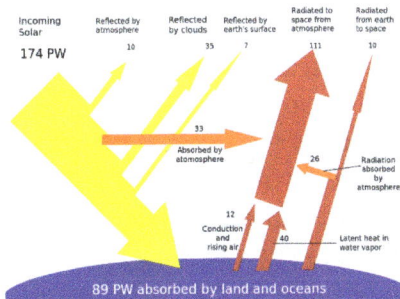

About half the incoming solar energy reaches the Earth's surface.

Solar radiation is absorbed by the Earth's land surface, oceans – which cover about 71% of the globe – and atmosphere. Warm air containing evaporated water from the oceans rises, causing atmospheric circulation or convection. When the air reaches a high altitude, where the temperature is low, water vapor condenses into clouds, which rain onto the Earth's surface, completing the water cycle. The latent heat of water condensation amplifies convection, producing atmospheric phenomena such as wind, cyclones and anti-cyclones. Sunlight absorbed by the oceans and land masses keeps the surface at an average temperature of 14 °C. By photosynthesis, green plants convert solar energy into chemically stored energy, which produces food, wood and the biomass from which fossil fuels are derived.

The total solar energy absorbed by Earth's atmosphere, oceans and land masses is approximately 3,850,000 exajoules (EJ) per year. In 2002, this was more energy in one hour than the world used in one year. Photosynthesis captures approximately 3,000 EJ per year in biomass. The amount of solar energy reaching the surface of the planet is so vast that in one year it is about twice as much as will ever be obtained from all of the Earth's non-renewable resources of coal, oil, natural gas, and mined uranium combined,

Yearly solar fluxes & human consumption[1]		
Solar	3,850,000	
Wind	2,250	
Biomass potential	~200	
Primary energy use[2]	539	
Electricity[2]	~67	
[1] Energy given in Exajoule (EJ) = 10^{18} J = 278 TWh [2] Consumption as of year 2010		

The potential solar energy that could be used by humans differs from the amount of solar energy present near the surface of the planet because factors such as geography, time variation, cloud cover, and the land available to humans limit the amount of solar energy that we can acquire.

Geography affects solar energy potential because areas that are closer to the equator have a greater amount of solar radiation. However, the use of photovoltaics that can follow the position of the sun can significantly increase the solar energy potential in areas that are farther from the equator. Time variation effects the potential of solar energy because during the nighttime there is little solar radiation on the surface of the Earth for solar panels to absorb. This limits the amount of energy that solar panels can absorb in one day. Cloud cover can affect the potential of solar panels because clouds block incoming light from the sun and reduce the light available for solar cells.

In addition, land availability has a large effect on the available solar energy because solar panels can only be set up on land that is otherwise unused and suitable for solar panels. Roofs have been found to be a suitable place for solar cells, as many people have discovered that they can collect energy directly from their homes this way. Other areas that are suitable for solar cells are lands that are not being used for businesses where solar plants can be established.

Solar technologies are characterized as either passive or active depending on the way they capture, convert and distribute sunlight and enable solar energy to be harnessed at different levels around the world, mostly depending on distance from the equator. Although solar energy refers primarily to the use of solar radiation for practical ends, all renewable energies, other than Geothermal power and Tidal power, derive their energy either directly or indirectly from the Sun.

Active solar techniques use photovoltaics, concentrated solar power, solar thermal collectors, pumps, and fans to convert sunlight into useful outputs. Passive solar techniques include selecting materials with favorable thermal properties, designing spaces that naturally circulate air, and referencing the position of a building to the Sun. Active solar technologies increase the supply of energy and are considered supply side technologies, while passive solar technologies reduce the need for alternate resources and are generally considered demand side technologies.

In 2000, the United Nations Development Programme, UN Department of Economic and Social Affairs, and World Energy Council published an estimate of the potential solar energy that could be used by humans each year that took into account factors such as insolation, cloud cover, and the land that is usable by humans. The estimate found that solar energy has a global potential of 1,575–49,837 EJ per year *(see table below)*.

Annual solar energy potential by region (Exajoules)										
Region	North America	Latin America and Caribbean	Western Europe	Central and Eastern Europe	Former Soviet Union	Middle East and North Africa	Sub-Saharan Africa	Pacific Asia	South Asia	Centrally planned Asia	Pacific OECD
Minimum	181.1	112.6	25.1	4.5	199.3	412.4	371.9	41.0	38.8	115.5	72.6
Maximum	7,410	3,385	914	154	8,655	11,060	9,528	994	1,339	4,135	2,263

Note:
- Total global annual solar energy potential amounts to 1,575 EJ (minimum) to 49,837 EJ (maximum)
- Data reflects assumptions of annual clear sky irradiance, annual average sky clearance, and available land area. All figures given in Exajoules.

Quantitative relation of global solar potential vs. the world's primary energy consumption:
- Ratio of potential vs. current consumption (402 EJ) as of year: 3.9 (minimum) to 124 (maximum)
- Ratio of potential vs. projected consumption by 2050 (590–1,050 EJ): 1.5–2.7 (minimum) to 47–84 (maximum)
- Ratio of potential vs. projected consumption by 2100 (880–1,900 EJ): 0.8–1.8 (minimum) to 26–57 (maximum)

Source: United Nations Development Programme – World Energy Assessment (2000)

Thermal Energy

Solar thermal technologies can be used for water heating, space heating, space cooling and process heat generation.

Early Commercial Adaptation

In 1897, Frank Shuman, a U.S. inventor, engineer and solar energy pioneer built a small demonstration solar engine that worked by reflecting solar energy onto square boxes filled with ether, which has a lower boiling point than water, and were fitted internally with black pipes which in turn powered a steam engine. In 1908 Shuman formed the Sun Power Company with the intent of building larger solar power plants. He, along with his technical advisor A.S.E. Ackermann and British physicist Sir Charles Vernon Boys, developed an improved system using mirrors to reflect solar energy upon collector boxes, increasing heating capacity to the extent that water could now be used instead of ether. Shuman then constructed a full-scale steam engine powered by low-pressure water, enabling him to patent the entire solar engine system by 1912.

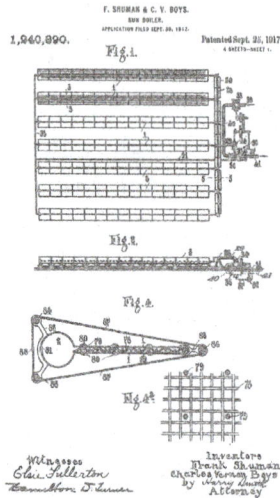

1917 Patent drawing of Shuman's solar collector

Shuman built the world's first solar thermal power station in Maadi, Egypt, between 1912 and 1913. His plant used parabolic troughs to power a 45–52 kilowatts (60–70 hp) engine that pumped more than 22,000 litres (4,800 imp gal; 5,800 US gal) of water per minute from the Nile River to adjacent cotton fields. Although the outbreak of World War I and the discovery of cheap oil in the 1930s discouraged the advancement of solar energy, Shuman's vision and basic design were resurrected in the 1970s with a new wave of interest in solar thermal energy. In 1916 Shuman was quoted in the media advocating solar energy's utilization, saying:

We have proved the commercial profit of sun power in the tropics and have more particularly proved that after our stores of oil and coal are exhausted the human race can receive unlimited power from the rays of the sun.

— Frank Shuman, New York Times, 2 July 1916

Water Heating

Solar hot water systems use sunlight to heat water. In low geographical latitudes (below 40 degrees) from 60 to 70% of the domestic hot water use with temperatures up to 60 °C can be provided by solar heating systems. The most common types of solar water heaters are evacuated tube collectors (44%) and glazed flat plate collectors (34%) generally used for domestic hot water; and unglazed plastic collectors (21%) used mainly to heat swimming pools.

Solar water heaters facing the Sun to maximize gain

As of 2007, the total installed capacity of solar hot water systems was approximately 154 thermal gigawatt (GW_{th}). China is the world leader in their deployment with 70 GW_{th} installed as of 2006 and a long-term goal of 210 GW_{th} by 2020. Israel and Cyprus are the per capita leaders in the use of solar hot water systems with over 90% of homes using them. In the United States, Canada, and Australia, heating swimming pools is the dominant application of solar hot water with an installed capacity of 18 GW_{th} as of 2005.

Heating, Cooling and Ventilation

In the United States, heating, ventilation and air conditioning (HVAC) systems account for 30% (4.65 EJ/yr) of the energy used in commercial buildings and nearly 50% (10.1 EJ/yr) of the energy used in residential buildings. Solar heating, cooling and ventilation technologies can be used to offset a portion of this energy.

MIT's Solar House #1, built in 1939 in the U.S., used seasonal thermal energy storage for year-round heating.

Thermal mass is any material that can be used to store heat—heat from the Sun in the case of solar energy. Common thermal mass materials include stone, cement and water. Historically they have been used in arid climates or warm temperate regions to keep buildings cool by absorbing solar energy during the day and radiating stored heat to the cooler atmosphere at night. However, they can be used in cold temperate areas to maintain warmth as well. The size and placement of thermal mass depend on several factors such as climate, daylighting and shading conditions. When properly incorporated, thermal mass maintains space temperatures in a comfortable range and reduces the need for auxiliary heating and cooling equipment.

A solar chimney (or thermal chimney, in this context) is a passive solar ventilation system composed of a vertical shaft connecting the interior and exterior of a building. As the chimney warms, the air inside is heated causing an updraft that pulls air through the building. Performance can be improved by using glazing and thermal mass materials in a way that mimics greenhouses.

Deciduous trees and plants have been promoted as a means of controlling solar heating and cooling. When planted on the southern side of a building in the northern hemisphere or the northern side in the southern hemisphere, their leaves provide shade during the summer, while the bare limbs allow light to pass during the winter. Since bare, leafless trees shade 1/3 to 1/2 of incident solar radiation, there is a balance between the benefits of summer shading and the corresponding loss of winter heating. In climates with significant heating loads, deciduous trees should not be planted on the Equator-facing side of a building because they will interfere with winter solar availability. They can, however, be used on the east and west sides to provide a degree of summer shading without appreciably affecting winter solar gain.

Cooking

Solar cookers use sunlight for cooking, drying and pasteurization. They can be grouped into three broad categories: box cookers, panel cookers and reflector cookers. The simplest solar cooker is the box cooker first built by Horace de Saussure in 1767. A basic box cooker consists of an insulated container with a transparent lid. It can be used effectively with partially overcast skies and will typically reach temperatures of 90–150 °C (194–302 °F). Panel cookers use a reflective panel to direct sunlight onto an insulated container and reach temperatures comparable to box cookers. Reflector cookers use various concentrating geometries (dish, trough, Fresnel mirrors) to focus light on a cooking container. These cookers reach temperatures of 315 °C (599 °F) and above but require direct light to function properly and must be repositioned to track the Sun.

Parabolic dish produces steam for cooking, in Auroville, India

Process Heat

Solar concentrating technologies such as parabolic dish, trough and Scheffler reflectors can provide process heat for commercial and industrial applications. The first commercial system was the Solar Total Energy Project (STEP) in Shenandoah, Georgia, USA where a field of 114 parabolic dishes provided 50% of the process heating, air conditioning and electrical requirements for a clothing factory. This grid-connected cogeneration system provided 400 kW of electricity plus thermal energy in the form of 401 kW steam and 468 kW chilled water, and had a one-hour peak load thermal storage. Evaporation ponds are shallow pools that concentrate dissolved solids through evaporation. The use of evaporation ponds to obtain salt from seawater is one of the oldest applications of solar energy. Modern uses include concentrating brine solutions used in leach mining and removing dissolved solids from waste streams. Clothes lines, clotheshorses, and clothes racks dry clothes through evaporation by wind and sunlight without consuming electricity or gas. In some states of the United States legislation protects the "right to dry" clothes. Unglazed transpired collectors (UTC) are perforated sun-facing walls used for preheating ventilation air. UTCs can raise the incoming air temperature up to 22 °C (40 °F) and deliver outlet temperatures of 45–60 °C (113–140 °F). The short payback period of transpired collectors (3 to 12 years) makes them a more cost-effective alternative than glazed collection systems. As of 2003, over 80 systems with a combined collector area of 35,000 square metres (380,000 sq ft) had been installed worldwide, including an 860 m² (9,300 sq ft) collector in Costa Rica used for drying coffee beans and a 1,300 m² (14,000 sq ft) collector in Coimbatore, India, used for drying marigolds.

Water Treatment

Solar water disinfection in Indonesia

Solar distillation can be used to make saline or brackish water potable. The first recorded instance of this was by 16th-century Arab alchemists. A large-scale solar distillation project was first constructed in 1872 in the Chilean mining town of Las Salinas. The plant, which had solar collection area of 4,700 m² (51,000 sq ft), could produce up to 22,700 L (5,000 imp gal; 6,000 US gal) per day and operate for 40 years. Individual still designs include single-slope, double-slope (or greenhouse type), vertical, conical, inverted absorber, multi-wick, and multiple effect. These stills can operate in passive, active, or hybrid modes. Double-slope stills are the most economical for decentralized domestic purposes, while active multiple effect units are more suitable for large-scale applications.

Solar water disinfection (SODIS) involves exposing water-filled plastic polyethylene terephthalate (PET) bottles to sunlight for several hours. Exposure times vary depending on weather and climate from a minimum of six hours to two days during fully overcast conditions. It is recommended by the World Health Organization as a viable method for household water treatment and safe storage. Over two million people in developing countries use this method for their daily drinking water.

Solar energy may be used in a water stabilization pond to treat waste water without chemicals or electricity. A further environmental advantage is that algae grow in such ponds and consume carbon dioxide in photosynthesis, although algae may produce toxic chemicals that make the water unusable.

Electricity Production

Solar power is the conversion of sunlight into electricity, either directly using photovoltaics (PV), or indirectly using concentrated solar power (CSP). CSP systems use lenses or mirrors and tracking systems to focus a large area of sunlight into a small beam. PV converts light into electric current using the photoelectric effect.

Some of the world's largest solar power stations: Ivanpah (CSP) and Topaz (PV)

Solar power is anticipated to become the world's largest source of electricity by 2050, with solar photovoltaics and concentrated solar power contributing 16 and 11 percent to the global overall consumption, respectively.

Commercial CSP plants were first developed in the 1980s. Since 1985 the eventually 354 MW SEGS CSP installation, in the Mojave Desert of California, is the largest solar power plant in the world. Other large CSP plants include the 150 MW Solnova Solar Power Station and the 100 MW Andasol solar power station, both in Spain. The 250 MW Agua Caliente Solar Project, in the United States, and the 221 MW Charanka Solar Park in India, are the world's largest photovoltaic plants. Solar projects exceeding 1 GW are being developed, but most of the deployed photovoltaics are in small rooftop arrays of less than 5 kW, which are connected to the grid using net metering and/or a feed-in tariff. In 2013 solar generated less than 1% of the world's total grid electricity.

Photovoltaics

In the last two decades, photovoltaics (PV), also known as solar PV, has evolved from a pure niche market of small scale applications towards becoming a mainstream electricity source. A solar cell is a device that converts light directly into electricity using the photoelectric effect. The first solar cell was constructed by Charles Fritts in the 1880s. In 1931 a German engineer, Dr Bruno Lange, developed a photo cell using silver selenide in place of copper oxide. Although the prototype selenium cells converted less than 1% of incident light into electricity, both Ernst Werner von Siemens and James Clerk Maxwell recognized the importance of this discovery. Following the work of Russell Ohl in the 1940s, researchers Gerald Pearson, Calvin Fuller and Daryl Chapin created the crystalline silicon solar cell in 1954. These early solar cells cost 286 USD/watt and reached efficiencies of 4.5–6%. By 2012 available efficiencies exceeded 20%, and the maximum efficiency of research photovoltaics was in excess of 40%.

Concentrated Solar Power

Concentrating Solar Power (CSP) systems use lenses or mirrors and tracking systems to focus a large area of sunlight into a small beam. The concentrated heat is then used as a heat source for a conventional power plant. A wide range of concentrating technologies exists; the most developed are the parabolic trough, the concentrating linear fresnel reflector, the Stirling dish and the solar power tower. Various techniques are used to track the Sun and focus light. In all of these systems a working fluid is heated by the concentrated sunlight, and is then used for power generation or energy storage.

Architecture and Urban Planning

Sunlight has influenced building design since the beginning of architectural history. Advanced solar architecture and urban planning methods were first employed by the Greeks and Chinese, who oriented their buildings toward the south to provide light and warmth.

The common features of passive solar architecture are orientation relative to the Sun, compact proportion (a low surface area to volume ratio), selective shading (overhangs) and thermal mass. When these features are tailored to the local climate and environment they can produce well-lit spaces that stay in a comfortable temperature range. Socrates' Megaron House is a classic example of passive solar design. The most recent approaches to solar design use computer modeling tying together solar lighting, heating and ventilation systems in an integrated solar design package. Active solar equipment such as pumps, fans and switchable windows can complement passive design and improve system performance.

Darmstadt University of Technology, Germany, won the 2007 Solar Decathlon in Washington, D.C. with this passive house designed for humid and hot subtropical climate.

Urban heat islands (UHI) are metropolitan areas with higher temperatures than that of the surrounding environment. The higher temperatures result from increased absorption of solar energy by urban materials such as asphalt and concrete, which have lower albedos and higher heat capacities than those in the natural environment. A straightforward method of counteracting the UHI effect is to paint buildings and roads white, and to plant trees in the area. Using these methods, a hypothetical "cool communities" program in Los Angeles has projected that urban temperatures could be reduced by approximately 3 °C at an estimated cost of US$1 billion, giving estimated total annual benefits of US$530 million from reduced air-conditioning costs and healthcare savings.

Agriculture and Horticulture

Greenhouses like these in the Westland municipality of the Netherlands grow vegetables, fruits and flowers.

Agriculture and horticulture seek to optimize the capture of solar energy in order to optimize the productivity of plants. Techniques such as timed planting cycles, tailored row orientation, staggered heights between rows and the mixing of plant varieties can improve crop yields. While sunlight is generally considered a plentiful resource, the exceptions highlight the importance of solar energy to agriculture. During the short growing seasons of the Little Ice Age, French and English farmers employed fruit walls to maximize the collection of solar energy. These walls acted as thermal masses and accelerated ripening by keeping plants warm. Early fruit walls were built perpendicular to the ground and facing south, but over time, sloping walls were developed to make better use of sunlight. In 1699, Nicolas Fatio de Duillier even suggested using a tracking mechanism which could pivot to follow the Sun. Applications of solar energy in agriculture aside from growing crops include pumping water, drying crops, brooding chicks and drying chicken manure. More recently the technology has been embraced by vintners, who use the energy generated by solar panels to power grape presses.

Greenhouses convert solar light to heat, enabling year-round production and the growth (in enclosed environments) of specialty crops and other plants not naturally suited to the local climate. Primitive greenhouses were first used during Roman times to produce cucumbers year-round for the Roman emperor Tiberius. The first modern greenhouses were built in Europe in the 16th century to keep exotic plants brought back from explorations abroad. Greenhouses remain an important part of horticulture today, and plastic transparent materials have also been used to similar effect in poly-tunnels and row covers.

Transport

Development of a solar-powered car has been an engineering goal since the 1980s. The World Solar Challenge is a biannual solar-powered car race, where teams from universities and enterprises compete over 3,021 kilometres (1,877 mi) across central Australia from Darwin to Adelaide. In 1987, when it was founded, the winner's average speed was 67 kilometres per hour (42 mph) and by 2007 the winner's average speed had improved to 90.87 kilometres per hour (56.46 mph). The North American Solar Challenge and the planned South African Solar Challenge are comparable competitions that reflect an international interest in the engineering and development of solar powered vehicles.

Winner of the 2013 World Solar Challenge in Australia

Some vehicles use solar panels for auxiliary power, such as for air conditioning, to keep the interior cool, thus reducing fuel consumption.

Solar electric aircraft circumnavigating the globe in 2015

In 1975, the first practical solar boat was constructed in England. By 1995, passenger boats incorporating PV panels began appearing and are now used extensively. In 1996, Kenichi Horie made the first solar-powered crossing of the Pacific Ocean, and the *Sun21* catamaran made the first solar-powered crossing of the Atlantic Ocean in the winter of 2006–2007. There were plans to circumnavigate the globe in 2010.

In 1974, the unmanned AstroFlight Sunrise airplane made the first solar flight. On 29 April 1979, the *Solar Riser* made the first flight in a solar-powered, fully controlled, man-carrying flying machine, reaching an altitude of 40 feet (12 m). In 1980, the *Gossamer Penguin* made the first piloted flights powered solely by photovoltaics. This was quickly followed by the *Solar Challenger* which crossed the English Channel in July 1981. In 1990 Eric Scott Raymond in 21 hops flew from California to North Carolina using solar power. Developments then turned back to unmanned aerial vehicles (UAV) with the *Pathfinder* (1997) and subsequent designs, culminating in the *Helios* which set the altitude record for a non-rocket-propelled aircraft at 29,524 metres (96,864 ft) in 2001. The *Zephyr*, developed by BAE Systems, is the latest in a line of record-breaking solar aircraft, making a 54-hour flight in 2007, and month-long flights were envisioned by 2010. As of 2016, Solar Impulse, an electric aircraft, is currently circumnavigating the globe. It is a single-seat plane powered by solar cells and capable of taking off under its own power. The design allows the aircraft to remain airborne for several days.

A solar balloon is a black balloon that is filled with ordinary air. As sunlight shines on the balloon, the air inside is heated and expands causing an upward buoyancy force, much like an artificially heated hot air balloon. Some solar balloons are large enough for human flight, but usage is generally limited to the toy market as the surface-area to payload-weight ratio is relatively high.

Fuel Production

Concentrated solar panels are getting a power boost. Pacific Northwest National Laboratory (PNNL) will be testing a new concentrated solar power system -- one that can help natural gas power plants reduce their fuel usage by up to 20 percent.

Solar chemical processes use solar energy to drive chemical reactions. These processes off-set energy that would otherwise come from a fossil fuel source and can also convert solar energy into storable and transportable fuels. Solar induced chemical reactions can be divided into thermochemical or photochemical. A variety of fuels can be produced by artificial photosynthesis. The multielectron catalytic chemistry involved in making carbon-based fuels (such as methanol) from reduction of carbon dioxide is challenging; a feasible alternative is hydrogen production from protons, though use of water as the source of electrons (as plants do) requires mastering the multielectron oxidation of two water molecules to molecular oxygen. Some have envisaged working solar fuel plants in coastal metropolitan areas by 2050 – the splitting of sea water providing hydrogen to be run through adjacent fuel-cell electric power plants and the pure water by-product going directly into the municipal water system. Another vision involves all human structures covering the earth's surface (i.e., roads, vehicles and buildings) doing photosynthesis more efficiently than plants.

Hydrogen production technologies have been a significant area of solar chemical research since the 1970s. Aside from electrolysis driven by photovoltaic or photochemical cells, several thermochemical processes have also been explored. One such route uses concentrators to split water into oxygen and hydrogen at high temperatures (2,300–2,600 °C or 4,200–4,700 °F). Another approach uses the heat from solar concentrators to drive the steam reformation of natural gas thereby increasing the overall hydrogen yield compared to conventional reforming methods. Thermochemical cycles characterized by the decomposition and regeneration of reactants present another avenue for hydrogen production. The Solzinc process under development at the Weizmann Institute of Science uses a 1 MW solar furnace to decompose zinc oxide (ZnO) at temperatures above 1,200 °C (2,200 °F). This initial reaction produces pure zinc, which can subsequently be reacted with water to produce hydrogen.

Energy Storage Methods

Thermal mass systems can store solar energy in the form of heat at domestically useful temperatures for daily or interseasonal durations. Thermal storage systems generally use readily available materials with high specific heat capacities such as water, earth and stone. Well-designed systems can lower peak demand, shift time-of-use to off-peak hours and reduce overall heating and cooling requirements.

Phase change materials such as paraffin wax and Glauber's salt are another thermal storage medium. These materials are inexpensive, readily available, and can deliver domestically useful temperatures (approximately 64 °C or 147 °F). The "Dover House" (in Dover, Massachusetts) was the first to use a Glauber's salt heating system, in 1948. Solar energy can also be stored at high temperatures using molten salts. Salts are an effective storage medium because they are low-cost, have a high specific heat capacity and can deliver heat at temperatures compatible with conventional power systems. The Solar Two project used this method of energy storage, allowing it to store 1.44 terajoules (400,000 kWh) in its 68 m³ storage tank with an annual storage efficiency of about 99%.

Thermal energy storage. The Andasol CSP plant uses tanks of molten salt to store solar energy.

Off-grid PV systems have traditionally used rechargeable batteries to store excess electricity. With grid-tied systems, excess electricity can be sent to the transmission grid, while standard grid electricity can be used to meet shortfalls. Net metering programs give household systems a credit for any electricity they deliver to the grid. This is handled by 'rolling back' the meter whenever the home produces more electricity than it consumes. If the net electricity use is below zero, the utility then rolls over the kilowatt hour credit to the next month. Other approaches involve the use of two meters, to measure electricity consumed vs. electricity produced. This is less common due to the increased installation cost of the second meter. Most standard meters accurately measure in both directions, making a second meter unnecessary.

Pumped-storage hydroelectricity stores energy in the form of water pumped when energy is available from a lower elevation reservoir to a higher elevation one. The energy is recovered when demand is high by releasing the water, with the pump becoming a hydroelectric power generator.

Development, Deployment and Economics

Beginning with the surge in coal use which accompanied the Industrial Revolution, energy consumption has steadily transitioned from wood and biomass to fossil fuels. The early development of solar technologies starting in the 1860s was driven by an expectation that coal would soon become scarce. However, development of solar technologies stagnated in the early 20th century in the face of the increasing availability, economy, and utility of coal and petroleum.

The 1973 oil embargo and 1979 energy crisis caused a reorganization of energy policies around the world and brought renewed attention to developing solar technologies. Deployment strategies focused on incentive programs such as the Federal Photovoltaic Utilization Program in the U.S. and the Sunshine Program in Japan. Other efforts included the formation of research facilities in the U.S. (SERI, now NREL), Japan (NEDO), and Germany (Fraunhofer Institute for Solar Energy Systems ISE).

Participants in a workshop on sustainable development inspect solar panels at Monterrey Institute of Technology and Higher Education, Mexico City on top of a building on campus.

Commercial solar water heaters began appearing in the United States in the 1890s. These systems saw increasing use until the 1920s but were gradually replaced by cheaper and more reliable heating fuels. As with photovoltaics, solar water heating attracted renewed attention as a result of the oil crises in the 1970s but interest subsided in the 1980s due to falling petroleum prices. Development in the solar water heating sector progressed steadily throughout the 1990s and annual growth rates have averaged 20% since 1999. Although generally underestimated, solar water heating and cooling is by far the most widely deployed solar technology with an estimated capacity of 154 GW as of 2007.

The International Energy Agency has said that solar energy can make considerable contributions to solving some of the most urgent problems the world now faces:

The development of affordable, inexhaustible and clean solar energy technologies will have huge longer-term benefits. It will increase countries' energy security through reliance on an indigenous, inexhaustible and mostly import-independent resource, enhance sustainability, reduce pollution, lower the costs of mitigating climate change, and keep fossil fuel prices lower than otherwise. These advantages are global. Hence the additional costs of the incentives for early deployment should be considered learning investments; they must be wisely spent and need to be widely shared.

In 2011, a report by the International Energy Agency found that solar energy technologies such as photovoltaics, solar hot water and concentrated solar power could provide a third of the world's energy by 2060 if politicians commit to limiting climate change. The energy from the sun could play a key role in de-carbonizing the global economy alongside improvements in energy efficiency and imposing costs on greenhouse gas emitters. "The strength of solar is the incredible variety and flexibility of applications, from small scale to big scale".

We have proved ... that after our stores of oil and coal are exhausted the human race can receive unlimited power from the rays of the sun.

— Frank Shuman, New York Times, 2 July 1916

ISO Standards

The International Organization for Standardization has established several standards relating to solar energy equipment. For example, ISO 9050 relates to glass in building while ISO 10217 relates to the materials used in solar water heaters.

Geothermal Energy

Geothermal energy is thermal energy generated and stored in the Earth. Thermal energy is the energy that determines the temperature of matter. The geothermal energy of the Earth's crust originates from the original formation of the planet and from radioactive decay of materials (in currently uncertain but possibly roughly equal proportions). The geothermal gradient, which is the difference in temperature between the core of the planet and its surface, drives a continuous conduction of thermal energy in the form of heat from the core to the surface.

Steam rising from the Nesjavellir Geothermal Power Station in Iceland.

Earth's internal heat is thermal energy generated from radioactive decay and continual heat loss from Earth's formation. Temperatures at the core–mantle boundary may reach over 4000 °C (7,200 °F). The high temperature and pressure in Earth's interior cause some rock to melt and solid mantle to behave plastically, resulting in portions of mantle convecting upward since it is lighter than the surrounding rock. Rock and water is heated in the crust, sometimes up to 370 °C (700 °F).

From hot springs, geothermal energy has been used for bathing since Paleolithic times and for space heating since ancient Roman times, but it is now better known for electricity generation. Worldwide, 11,700 megawatts (MW) of geothermal power is online in 2013. An additional 28 gigawatts of direct geothermal heating capacity is installed for district heating, space heating, spas, industrial processes, desalination and agricultural applications in 2010.

Geothermal power is cost-effective, reliable, sustainable, and environmentally friendly, but has historically been limited to areas near tectonic plate boundaries. Recent technological advances have dramatically expanded the range and size of viable resources, especially for applications such as home heating, opening a potential for widespread exploitation. Geothermal wells release greenhouse gases trapped deep within the earth, but these emissions are much lower per energy unit than those of fossil fuels. As a result, geothermal power has the potential to help mitigate global warming if widely deployed in place of fossil fuels.

The Earth's geothermal resources are theoretically more than adequate to supply humanity's energy needs, but only a very small fraction may be profitably exploited. Drilling and exploration for deep resources is very expensive. Forecasts for the future of geothermal power depend on assumptions about technology, energy prices, subsidies, and interest rates. Pilot programs like EWEB's customer opt in Green Power Program show that customers would be willing to pay a little more for a renewable energy source like geothermal. But as a result of government assisted research and industry experience, the cost of generating geothermal power has decreased by 25% over the past two decades. In 2001, geothermal energy costs between two and ten US cents per kWh.

History

Hot springs have been used for bathing at least since Paleolithic times. The oldest known spa is a stone pool on China's Lisan mountain built in the Qin Dynasty in the 3rd century BC, at the same site where the Huaqing Chi palace was later built. In the first century AD, Romans conquered *Aquae Sulis*, now Bath, Somerset, England, and used the hot springs there to feed public baths and underfloor heating. The admission fees for these baths probably represent the first commercial use of geothermal power. The world's oldest geothermal district heating system in Chaudes-Aigues, France, has been operating since the 14th century. The earliest industrial exploitation began in 1827 with the use of geyser steam to extract boric acid from volcanic mud in Larderello, Italy.

The oldest known pool fed by a hot spring, built in the Qin dynasty in the 3rd century BCE.

In 1892, America's first district heating system in Boise, Idaho was powered directly by geothermal energy, and was copied in Klamath Falls, Oregon in 1900. A deep geothermal well was used to heat greenhouses in Boise in 1926, and geysers were used to heat greenhouses in Iceland and Tuscany at about the same time. Charlie Lieb developed the first downhole heat exchanger in 1930 to heat his house. Steam and hot water from geysers began heating homes in Iceland starting in 1943.

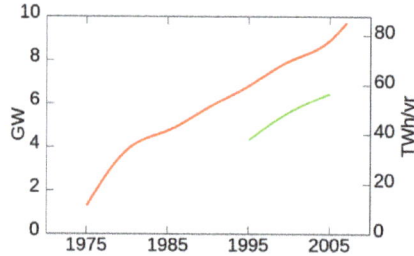

Global geothermal electric capacity. Upper red line is installed capacity; lower green line is realized production.

In the 20th century, demand for electricity led to the consideration of geothermal power as a generating source. Prince Piero Ginori Conti tested the first geothermal power generator on 4 July 1904, at the same Larderello dry steam field where geothermal acid extraction began. It successfully lit four light bulbs. Later, in 1911, the world's first commercial geothermal power plant was built there. It was the world's only industrial producer of geothermal electricity until New Zealand built a plant in 1958. In 2012, it produced some 594 megawatts.

Lord Kelvin invented the heat pump in 1852, and Heinrich Zoelly had patented the idea of using it to draw heat from the ground in 1912. But it was not until the late 1940s that the geothermal heat pump was successfully implemented. The earliest one was probably Robert C. Webber's home-made 2.2 kW direct-exchange system, but sources disagree as to the exact timeline of his invention. J. Donald Kroeker designed the first commercial geothermal heat pump to heat the Commonwealth Building (Portland, Oregon) and demonstrated it in 1946. Professor Carl Nielsen of Ohio State University built the first residential open loop version in his home in 1948. The technology became popular in Sweden as a result of the 1973 oil crisis, and has been growing slowly in worldwide acceptance since then. The 1979 development of polybutylene pipe greatly augmented the heat pump's economic viability.

In 1960, Pacific Gas and Electric began operation of the first successful geothermal electric power plant in the United States at The Geysers in California. The original turbine lasted for more than 30 years and produced 11 MW net power.

The binary cycle power plant was first demonstrated in 1967 in the USSR and later introduced to the US in 1981. This technology allows the generation of electricity from much lower temperature resources than previously. In 2006, a binary cycle plant in Chena Hot Springs, Alaska, came on-line, producing electricity from a record low fluid temperature of 57 °C (135 °F).

Direct Usage

Direct Use Data 2015	
Country	**Usage (MWt) 2015**
United States	17,415.91
Philippines	3.30
Indonesia	2.30
Mexico	155.82
Italy	1,014.00
New Zealand	487.45
Iceland	2,040.00
Japan	2,186.17
Iran	81.50
El Salvador	3.36
Kenya	22.40
Costa Rica	1.00
Russia	308.20
Turkey	2,886.30
Papua-New Guinea	0.10
Guatemala	2.31
Portugal	35.20
China	17,870.00
France	2,346.90
Ethiopia	2.20
Germany	2,848.60
Austria	903.40
Australia	16.09
Thailand	128.51

Electricity

The International Geothermal Association (IGA) has reported that 10,715 megawatts (MW) of geothermal power in 24 countries is online, which was expected to generate 67,246 GWh of electricity in 2010. This represents a 20% increase in online capacity since 2005. IGA projects growth to 18,500 MW by 2015, due to the projects presently under consideration, often in areas previously assumed to have little exploitable resource.

In 2010, the United States led the world in geothermal electricity production with 3,086 MW of installed capacity from 77 power plants. The largest group of geothermal power plants in the world is located at The Geysers, a geothermal field in California. The Philippines is the second highest producer, with 1,904 MW of capacity online. Geothermal power makes up approximately 27% of Philippine electricity generation.

Installed geothermal electric capacity				
Country	Capacity (MW) 2007	Capacity (MW) 2010	Percentage of national electricity production	Percentage of global geothermal production
United States	2687	3086	0.3	29
Philippines	1969.7	1904	27	18
Indonesia	992	1197	3.7	11
Mexico	953	958	3	9
Italy	810.5	843	1.5	8
New Zealand	471.6	628	10	6
Iceland	421.2	575	30	5
Japan	535.2	536	0.1	5
Iran	250	250		
El Salvador	204.2	204	25	
Kenya	128.8	167	11.2	
Costa Rica	162.5	166	14	
Nicaragua	87.4	88	10	
Russia	79	82		
Turkey	38	82		
Papua-New Guinea	56	56		
Guatemala	53	52		
Portugal	23	29		
China	27.8	24		
France	14.7	16		
Ethiopia	7.3	7.3		
Germany	8.4	6.6		
Austria	1.1	1.4		
Australia	0.2	1.1		
Thailand	0.3	0.3		
TOTAL	**9,981.9**	**10,959.7**		

Geothermal electric plants were traditionally built exclusively on the edges of tectonic plates where high temperature geothermal resources are available near the surface. The development of binary cycle power plants and improvements in drilling and extraction technology enable enhanced geothermal systems over a much greater geographical range. Demonstration projects are operational in Landau-Pfalz, Germany, and Soultz-sous-Forêts, France, while an earlier effort in Basel, Switzerland was shut down after it triggered earthquakes. Other demonstration projects are under construction in Australia, the United Kingdom, and the United States of America.

The thermal efficiency of geothermal electric plants is low, around 10–23%, because geothermal fluids do not reach the high temperatures of steam from boilers. The laws of thermodynamics limits the efficiency of heat engines in extracting useful energy. Exhaust heat is wasted, unless it can be used directly and locally, for example in greenhouses, timber mills, and district heating. System efficiency does not materially affect operational costs as it would for plants that use fuel, but it does affect return on the capital used to build the plant. In order to produce more energy than the pumps consume, electricity generation requires relatively hot fields and specialized heat cycles. Because geothermal power does not rely on variable sources of energy, unlike, for example, wind or solar, its capacity factor can be quite large – up to 96% has been demonstrated. The global average was 73% in 2005.

Types

Geothermal energy comes in either *vapor-dominated* or *liquid-dominated* forms. Larderello and The Geysers are vapor-dominated. Vapor-dominated sites offer temperatures from 240 to 300 °C that produce superheated steam.

Liquid-dominated Plants

Liquid-dominated reservoirs (LDRs) were more common with temperatures greater than 200 °C (392 °F) and are found near young volcanoes surrounding the Pacific Ocean and in rift zones and hot spots. *Flash plants* are the common way to generate electricity from these sources. Pumps are generally not required, powered instead when the water turns to steam. Most wells generate 2-10MWe. Steam is separated from liquid via cyclone separators, while the liquid is returned to the reservoir for reheating/reuse. As of 2013, the largest liquid system is Cerro Prieto in Mexico, which generates 750 MWe from temperatures reaching 350 °C (662 °F). The Salton Sea field in Southern California offers the potential of generating 2000 MWe.

Lower temperature LDRs (120–200 °C) require pumping. They are common in extensional terrains, where heating takes place via deep circulation along faults, such as in the Western US and Turkey. Water passes through a heat exchanger in a Rankine cycle binary plant. The water vaporizes an organic working fluid that drives a turbine. These binary plants originated in the Soviet Union in the late 1960s and predominate in new US plants. Binary plants have no emissions.

Thermal Energy

Lower temperature sources produce the energy equivalent of 100M BBL per year. Sources with temperatures of 30–150 °C are used without conversion to electricity as district heating, greenhouses, fisheries, mineral recovery, industrial process heating and bathing in 75 countries. Heat pumps extract energy from shallow sources at 10–20 °C in 43 countries for use in space heating and cooling. Home heating is the fastest-growing means of exploiting geothermal energy, with global annual growth rate of 30% in 2005 and 20% in 2012.

Approximately 270 petajoules (PJ) of geothermal heating was used in 2004. More than half went for space heating, and another third for heated pools. The remainder supported industrial and agricultural applications. Global installed capacity was 28 GW, but capacity factors tend to be low (30% on average) since heat is mostly needed in winter. Some 88 PJ for space heating was extracted by an estimated 1.3 million geothermal heat pumps with a total capacity of 15 GW.

Heat for these purposes may also be extracted from co-generation at a geothermal electrical plant.

Heating is cost-effective at many more sites than electricity generation. At natural hot springs or geysers, water can be piped directly into radiators. In hot, dry ground, earth tubes or downhole heat exchangers can collect the heat. However, even in areas where the ground is colder than room temperature, heat can often be extracted with a geothermal heat pump more cost-effectively and cleanly than by conventional furnaces. These devices draw on much shallower and colder resources than traditional geothermal techniques. They frequently combine functions, including air conditioning, seasonal thermal energy storage, solar energy collection, and electric heating. Heat pumps can be used for space heating essentially anywhere.

Iceland is the world leader in direct applications. Some 92.5% of its homes are heated with geothermal energy, saving Iceland over $100 million annually in avoided oil imports. Reykjavík, Iceland has the world's biggest district heating system. Once known as the most polluted city in the world, it is now one of the cleanest.

Enhanced Geothermal

Enhanced geothermal systems (EGS) actively inject water into wells to be heated and pumped back out. The water is injected under high pressure to expand existing rock fissures to enable the water to freely flow in and out. The technique was adapted from oil and gas extraction techniques. However, the geologic formations are deeper and no toxic chemicals are used, reducing the possibility of environmental damage. Drillers can employ directional drilling to expand the size of the reservoir.

Small-scale EGS have been installed in the Rhine Graben at Soultz-sous-Forêts in France and at Landau and Insheim in Germany.

Economics

Geothermal power requires no fuel (except for pumps), and is therefore immune to fuel cost fluctuations. However, capital costs are significant. Drilling accounts for over half the costs, and exploration of deep resources entails significant risks. A typical well doublet (extraction and injection wells) in Nevada can support 4.5 megawatts (MW) and costs about $10 million to drill, with a 20% failure rate.

A power plant at The Geysers

In total, electrical plant construction and well drilling cost about €2–5 million per MW of electrical capacity, while the break–even price is 0.04–0.10 € per kW·h. Enhanced geothermal systems tend to be on the high side of these ranges, with capital costs above $4 million per MW and break–even above $0.054 per kW·h in 2007. Direct heating applications can use much shallower wells with lower temperatures, so smaller systems with lower costs and risks are feasible. Residential geothermal heat pumps with a capacity of 10 kilowatt (kW) are routinely installed for around $1–3,000 per kilowatt. District heating systems may benefit from economies of scale if demand is geographically dense, as in cities and greenhouses, but otherwise piping installation dominates capital costs. The capital cost of one such district heating system in Bavaria was estimated at somewhat over 1 million € per MW. Direct systems of any size are much simpler than electric generators and have lower maintenance costs per kW·h, but they must consume electricity to run pumps and compressors. Some governments subsidize geothermal projects.

Geothermal power is highly scalable: from a rural village to an entire city.

The most developed geothermal field in the United States is The Geysers in Northern California.

Geothermal projects have several stages of development. Each phase has associated risks. At the early stages of reconnaissance and geophysical surveys, many projects are cancelled, making that phase unsuitable for traditional lending. Projects moving forward from the identification, exploration and exploratory drilling often trade equity for financing.

Resources

The Earth's internal thermal energy flows to the surface by conduction at a rate of 44.2 terawatts (TW), and is replenished by radioactive decay of minerals at a rate of 30 TW. These power rates are more than double humanity's current energy consumption from all primary sources, but most of this energy flow is not recoverable. In addition to the internal heat flows, the top layer of the surface to a depth of 10 meters (33 ft) is heated by solar energy during the summer, and releases that energy and cools during the winter.

Outside of the seasonal variations, the geothermal gradient of temperatures through the crust is 25–30 °C (77–86 °F) per kilometer of depth in most of the world. The conductive heat flux averages 0.1 MW/km². These values are much higher near tectonic plate boundaries where the crust is thinner. They may be further augmented by fluid circulation, either through magma conduits, hot springs, hydrothermal circulation or a combination of these.

A geothermal heat pump can extract enough heat from shallow ground anywhere in the world to provide home heating, but industrial applications need the higher temperatures of deep resources. The thermal efficiency and profitability of electricity generation is particularly sensitive to temperature. The most demanding applications receive the greatest benefit from a high natural heat flux, ideally from using a hot spring. The next best option is to drill a well into a hot aquifer. If no adequate aquifer is available, an artificial one may be built by injecting water to hydraulically fracture the bedrock. This last approach is called hot dry rock geothermal energy in Europe, or enhanced geothermal systems in North America. Much greater potential may be available from this approach than from conventional tapping of natural aquifers.

Estimates of the potential for electricity generation from geothermal energy vary sixfold, from .035to2TW depending on the scale of investments. Upper estimates of geothermal resources assume enhanced geothermal wells as deep as 10 kilometres (6 mi), whereas existing geothermal wells are rarely more than 3 kilometres (2 mi) deep. Wells of this depth are now common in the petroleum industry. The deepest research well in the world, the Kola superdeep borehole, is 12 kilometres (7 mi) deep.

Production

According to the Geothermal Energy Association (GEA) installed geothermal capacity in the United States grew by 5%, or 147.05 MW, ce the last annual survey in March 2012. This increase came from seven geothermal projects that began production in 2012. GEA also revised its 2011 estimate of installed capacity upward by 128 MW, bringing current installed U.S. geothermal capacity to 3,386 MW.

Myanmar Engineering Society has identified at least 39 locations capable of geothermal power production and some of these hydrothermal reservoirs lie quite close to Yangon which is a significant underutilized resource.

Renewability and Sustainability

Geothermal power is considered to be renewable because any projected heat extraction is small compared to the Earth's heat content. The Earth has an internal heat content of 10^{31} joules ($3 \cdot 10^{15}$ TW·hr), approximately 100 billion times current (2010) worldwide annual energy consumption. About 20% of this is residual heat from planetary accretion, and the remainder is attributed to higher radioactive decay rates that existed in the past. Natural heat flows are not in equilibrium, and the planet is slowly cooling down on geologic timescales. Human extraction taps a minute fraction of the natural outflow, often without accelerating it.

Geothermal power is also considered to be sustainable thanks to its power to sustain the Earth's intricate ecosystems. By using geothermal sources of energy present generations of humans will not endanger the capability of future generations to use their own resources to the same amount that those energy sources are presently used. Further, due to its low emissions geothermal energy is considered to have excellent potential for mitigation of global warming.

Even though geothermal power is globally sustainable, extraction must still be monitored to avoid local depletion. Over the course of decades, individual wells draw down local temperatures and water levels until a new equilibrium is reached with natural flows. The three oldest sites, at Larderello, Wairakei, and the Geysers have experienced reduced output because of local depletion. Heat and water, in uncertain proportions, were extracted faster than they were replenished. If production is reduced and water is reinjected, these wells could theoretically recover their full potential. Such mitigation strategies have already been implemented at some sites. The long-term sustainability of geothermal energy has been demonstrated at the Lardarello field in Italy since 1913, at the Wairakei field in New Zealand since 1958, and at The Geysers field in California since 1960.

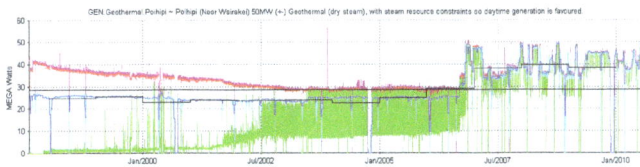

Electricity Generation at Poihipi, New Zealand.

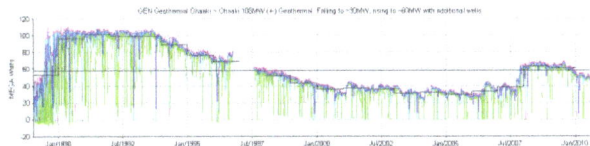

Electricity Generation at Ohaaki, New Zealand.

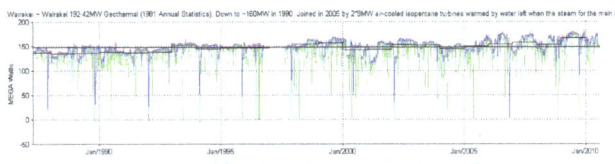

Electricity Generation at Wairakei, New Zealand.

Falling electricity production may be boosted through drilling additional supply boreholes, as at Poihipi and Ohaaki. The Wairakei power station has been running much longer, with its first unit commissioned in November 1958, and it attained its peak generation of 173MW in 1965, but already the supply of high-pressure steam was faltering, in 1982 being derated to intermediate pressure and the station managing 157MW. Around the start of the 21st century it was managing about 150MW, then in 2005 two 8MW isopentane systems were added, boosting the station's output by about 14MW. Detailed data are unavailable, being lost due to re-organisations. One such re-organisation in 1996 causes the absence of early data for Poihipi (started 1996), and the gap in 1996/7 for Wairakei and Ohaaki; half-hourly data for Ohaaki's first few months of operation are also missing, as well as for most of Wairakei's history.

Environmental Effects

Fluids drawn from the deep earth carry a mixture of gases, notably carbon dioxide (CO_2), hydrogen sulfide (H_2S), methane (CH_4) and ammonia (NH_3). These pollutants contribute to global warming, acid rain, and noxious smells if released. Existing geothermal electric plants emit an average of 122 kilograms (269 lb) of CO_2 per megawatt-hour (MW·h) of electricity, a small fraction of the emission intensity of conventional fossil fuel plants. Plants that experience high levels of acids and volatile chemicals are usually equipped with emission-control systems to reduce the exhaust.

Krafla Geothermal Station in northeast Iceland

Geothermal power station in the Philippines

In addition to dissolved gases, hot water from geothermal sources may hold in solution trace amounts of toxic elements such as mercury, arsenic, boron, and antimony. These chemicals precipitate as the water cools, and can cause environmental damage if released. The modern practice of injecting cooled geothermal fluids back into the Earth to stimulate production has the side benefit of reducing this environmental risk.

Direct geothermal heating systems contain pumps and compressors, which may consume energy from a polluting source. This parasitic load is normally a fraction of the heat output, so it is always less polluting than electric heating. However, if the electricity is produced by burning fossil fuels, then the net emissions of geothermal heating may be comparable to directly burning the fuel for heat. For example, a geothermal heat pump powered by electricity from a combined cycle natural gas plant would produce about as much pollution as a natural gas condensing furnace of the same size. Therefore, the environmental value of direct geothermal heating applications is highly dependent on the emissions intensity of the neighboring electric grid.

Plant construction can adversely affect land stability. Subsidence has occurred in the Wairakei field in New Zealand. In Staufen im Breisgau, Germany, tectonic uplift occurred instead, due to a previously isolated anhydrite layer coming in contact with water and turning into gypsum, doubling its volume. Enhanced geothermal systems can trigger earthquakes as part of hydraulic fracturing. The project in Basel, Switzerland was suspended because more than 10,000 seismic events measuring up to 3.4 on the Richter Scale occurred over the first 6 days of water injection.

Geothermal has minimal land and freshwater requirements. Geothermal plants use 3.5 square kilometres (1.4 sq mi) per gigawatt of electrical production (not capacity) versus 32 square kilometres (12 sq mi) and 12 square kilometres (4.6 sq mi) for coal facilities and wind farms respectively. They use 20 litres (5.3 US gal) of freshwater per MW·h versus over 1,000 litres (260 US gal) per MW·h for nuclear, coal, or oil.

Legal Frameworks

Some of the legal issues raised by geothermal energy resources include questions of ownership and allocation of the resource, the grant of exploration permits, exploitation rights, royalties, and the extent to which geothermal energy issues have been recognized in existing planning and environmental laws. Other questions concern overlap between geothermal and mineral or petroleum tenements. Broader issues concern the extent to which the legal framework for encouragement of renewable energy assists in encouraging geothermal industry innovation and development.

References

- Spellman, Frank R. (2013). Safe Work Practices for Green Energy Jobs (first ed.). DEStech Publications. p. 323. ISBN 978-1-60595-075-4. Retrieved 29 December 2014.

- History of Wind Energy in Cutler J. Cleveland,(ed) Encyclopedia of Energy Vol.6, Elsevier, ISBN 978-1-60119-433-6.

- Turcotte, D. L.; Schubert, G. (2002). "Chapter 4". Geodynamics (2nd ed.). Cambridge, England, UK: Cambridge University Press. pp. 136–137. ISBN 978-0-521-66624-4.

- Anderson, Lorraine; Palkovic, Rick (1994). Cooking with Sunshine (The Complete Guide to Solar Cuisine with 150 Easy Sun-Cooked Recipes). Marlowe & Company. ISBN 1-56924-300-X.

- Bradford, Travis (2006). Solar Revolution: The Economic Transformation of the Global Energy Industry. MIT Press. ISBN 0-262-02604-X.

- Butti, Ken; Perlin, John (1981). A Golden Thread (2500 Years of Solar Architecture and Technology). Van Nostrand Reinhold. ISBN 0-442-24005-8.

- Martin, Christopher L.; Goswami, D. Yogi (2005). Solar Energy Pocket Reference. International Solar Energy Society. ISBN 0-9771282-0-2.

- Scheer, Hermann (2002). The Solar Economy (Renewable Energy for a Sustainable Global Future). Earthscan Publications Ltd. ISBN 1-84407-075-1.

- Schittich, Christian (2003). Solar Architecture (Strategies Visions Concepts). Architektur-Dokumentation GmbH & Co. KG. ISBN 3-7643-0747-1.

- Smil, Vaclav (2003). Energy at the Crossroads: Global Perspectives and Uncertainties. MIT Press. p. 443. ISBN 0-262-19492-9.

- Smil, Vaclav (2006). Energy at the Crossroads (PDF). Organisation for Economic Co-operation and Development. ISBN 0-262-19492-9. Retrieved 29 September 2007.

- Turcotte, D. L.; Schubert, G. (2002), Geodynamics (2 ed.), Cambridge, England, UK: Cambridge University Press, pp. 136–137, ISBN 978-0-521-66624-4 .

Biomass: An Integrated Study

Biomass is the organic matter that is derived from living organisms. It usually refers to plants or plant-based materials. The following text focuses on topics such as renewable natural gas and biochar. The topics discussed in the chapter are of great importance to broaden the existing knowledge on biomass.

Biomass

Biomass is organic matter derived from living, or recently living organisms. Biomass can be used as a source of energy and it most often refers to plants or plant-based materials which are not used for food or feed, and are specifically called lignocellulosic biomass. As an energy source, biomass can either be used directly via combustion to produce heat, or indirectly after converting it to various forms of biofuel. Conversion of biomass to biofuel can be achieved by different methods which are broadly classified into: *thermal*, *chemical*, and *biochemical* methods.

Biomass Sources

Historically, humans have harnessed biomass-derived energy since the time when people began burning wood to make fire. Even today, biomass is the only source of fuel for domestic use in many developing countries. Biomass is all biologically-produced matter based in carbon, hydrogen and oxygen. The estimated biomass production in the world is 104.9 petagrams (104.9 × 10^{15} g - about 105 billion metric tons) of carbon per year, about half in the ocean and half on land.

Wood remains the largest biomass energy source today; examples include forest residues (such as dead trees, branches and tree stumps), yard clippings, wood chips and even municipal solid waste. Wood energy is derived by using lignocellulosic biomass (second-generation biofuels) as fuel. Harvested wood may be used directly as a fuel or collected from wood waste streams. The largest source of energy from wood is pulping liquor or "black liquor," a waste product from processes of the pulp, paper and paperboard industry. In the second sense, biomass includes plant or animal matter that can be converted into fibers or other industrial chemicals, including biofuels. Industrial biomass can be grown from numerous types of plants, including miscanthus, switchgrass, hemp, corn, poplar, willow, sorghum, sugarcane, bamboo, and a variety of tree species, ranging from eucalyptus to oil palm (palm oil).

Eucalyptus in Brazil. Remains of the tree are reused for power generation.

Based on the source of biomass, biofuels are classified broadly into two major categories. First-generation biofuels are derived from sources such as sugarcane and corn starch. Sugars present in this biomass are fermented to produce bioethanol, an alcohol fuel which can be used directly in a fuel cell to produce electricity or serve as an additive to gasoline. However, utilizing food-based resources for fuel production only aggravates the food shortage problem. Second-generation biofuels, on the other hand, utilize non-food-based biomass sources such as agriculture and municipal waste. These biofuels mostly consist of lignocellulosic biomass, which is not edible and is a low-value waste for many industries. Despite being the favored alternative, economical production of second-generation biofuel is not yet achieved due to technological issues. These issues arise mainly due to chemical inertness and structural rigidity of lignocellulosic biomass.

Plant energy is produced by crops specifically grown for use as fuel that offer high biomass output per hectare with low input energy. Some examples of these plants are wheat, which typically yields 7.5–8 tonnes of grain per hectare, and straw, which typically yields 3.5–5 tonnes per hectare in the UK. The grain can be used for liquid transportation fuels while the straw can be burned to produce heat or electricity. Plant biomass can also be degraded from cellulose to glucose through a series of chemical treatments, and the resulting sugar can then be used as a first-generation biofuel.

The main contributors of waste energy are municipal solid waste, manufacturing waste, and landfill gas. Energy derived from biomass is projected to be the largest non-hydro-electric renewable resource of electricity in the US between 2000 and 2020.

Biomass can be converted to other usable forms of energy like methane gas or transportation fuels like ethanol and biodiesel. Rotting garbage, and agricultural and human waste, all release methane gas, also called landfill gas or biogas. Crops such as corn and sugarcane can be fermented to produce the transportation fuel ethanol. Biodiesel, another transportation fuel, can be produced from leftover food products like vegetable oils and animal fats. Also, biomass-to-liquids (called "BTLs") and cellulosic ethanol are still under research.

There is research involving algae, or algae-derived, biomass, as this non-food resource can be produced at rates five to ten times those of other types of land-based agriculture, such as corn and soy. Once harvested, it can be fermented to produce biofuels such as ethanol, butanol, and methane, as well as biodiesel and hydrogen. Efforts are being made to identify which species of algae are most suitable for energy production. Genetic engineering approaches could also be utilized to improve microalgae as a source of biofuel.

The biomass used for electricity generation varies by region. Forest by-products, such as wood residues, are common in the US. Agricultural waste is common in Mauritius (sugar cane residue) and Southeast Asia (rice husks). Animal husbandry residues, such as poultry litter, are common in the UK.

As of 2015, a new bioenergy sewage treatment process aimed at developing countries is under trial; the Omni Processor is a self-sustaining process which uses sewerage solids as fuel in a process to convert waste water into drinking water, with surplus electrical energy being generated for export.

Comparison of Total Plant Biomass Yields (Dry Basis)

World Resources

If the total annual primary production of biomass is just over 100 billion (1.0E+11) tonnes of Carbon /yr, and the energy reserve per metric tonne of biomass is between about 1.5E3 – 3E3 Kilowatt hours (5E6 – 10E6 BTU), or 24.8 TW average, then biomass could in principle provide 1.4 times the approximate annual 150E3 Terrawatt.hours required for the current world energy consumption. For reference, the total solar power on Earth is 174 kTW. The biomass equivalent to solar energy ratio is 143 ppm (parts per million), given current living system coverage on Earth. Best in class solar cell efficiency is (20-40)%. Additionally, Earth's internal radioactive energy production, largely the driver for volcanic activity, continental drift, etc., is in the same range of power, 20 TW. At some 50% carbon mass content in biomass, annual production, this corresponds to about 6% of atmospheric carbon content in CO_2 (for the current 400 ppm).

> (1.0E+11 tonnes biomass annually produced approximately 25 TW)
>
> Annual world biomass energy equivalent =16.7 - 33.4 TW.
>
> Annual world energy consumption =17.7. On average, biomass production is 1.4 times larger than world energy consumption.

Common Commodity Food Crops

- Agave: 1–21 tons/acre

- Alfalfa: 4–6 tons/acre

- Barley: grains – 1.6–2.8 tons/acre, straw – 0.9–2.5 tons/acre, total – 2.5–5.3 tons/acre

- Corn: grains – 3.2–4.9 tons/acre, stalks and stovers – 2.3–3.4 tons/acre, total – 5.5–8.3 tons/acre

- Jerusalem artichokes: tubers 1–8 tons/acre, tops 2–13 tons/acre, total 9–13 tons/acre

- Oats: grains – 1.4–5.4 tons/acre, straw – 1.9–3.2 tons/acre, total – 3.3–8.6 tons/acre

- Rye: grains – 2.1–2.4 tons/acre, straw – 2.4–3.4 tons/acre, total – 4.5–5.8 tons/acre

- Wheat: grains – 1.2–4.1 tons/acre, straw – 1.6–3.8 tons/acre, total – 2.8–7.9 tons/acre

Woody Crops

- Oil palm: fronds 11 ton/acre, whole fruit bunches 1 ton/acre, trunks 30 ton/acre

Not Yet in Commercial Planting

- Giant miscanthus: 5–15 tons/acre

- Sunn hemp: 4.5 tons/acre

- Switchgrass: 4–6 tons/acre

Genetically Modified Varieties

- Energy Sorghum

Biomass Conversion

Thermal Conversion

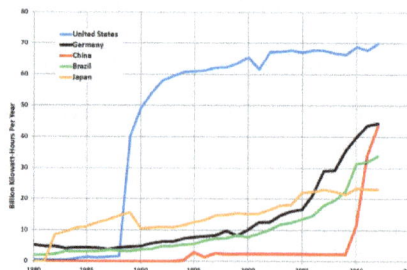

Trends in the top five countries generating electricity from biomass

Thermal conversion processes use heat as the dominant mechanism to convert biomass into another chemical form. The basic alternatives of combustion (torrefaction, pyrolysis, and gasification) are separated principally by the extent to which the chemical reactions involved are allowed to proceed (mainly controlled by the availability of oxygen and conversion temperature).

Energy created by burning biomass (fuel wood) is particularly suited for countries where the fuel wood grows more rapidly, e.g. tropical countries. There are a number of other less common, more experimental or proprietary thermal processes that may offer benefits such as hydrothermal upgrading (HTU) and hydroprocessing. Some have been developed for use on high moisture content biomass, including aqueous slurries, and allow them to be converted into more convenient forms. Some of the applications of thermal conversion are combined heat and power (CHP) and co-firing. In a typical dedicated biomass power plant, efficiencies range from 20–27% (higher heating value basis). Biomass cofiring with coal, by contrast, typically occurs at efficiencies near those of the coal combustor (30–40%, higher heating value basis).

Chemical Conversion

A range of chemical processes may be used to convert biomass into other forms, such as to produce a fuel that is more conveniently used, transported or stored, or to exploit some property of the process itself. Many of these processes are based in large part on similar coal-based processes, such as Fischer-Tropsch synthesis, methanol production, olefins (ethylene and propylene), and similar chemical or fuel feedstocks. In most cases, the first step involves gasification, which step generally is the most expensive and involves the greatest technical risk. Biomass is more difficult to feed into a pressure vessel than coal or any liquid. Therefore, biomass gasification is frequently done at atmospheric pressure and causes combustion of biomass to produce a combustible gas consisting of carbon monoxide, hydrogen, and traces of methane. This gas mixture, called a producer gas, can provide fuel for various vital processes, such as internal combustion engines, as well as substitute for furnace oil in direct heat applications. Because any biomass material can undergo gasification, this process is far more attractive than ethanol or biomass production, where only particular biomass materials can be used to produce a fuel. In addition, biomass gasification is a desirable process due to the ease at which it can convert solid waste (such as wastes available on a farm) into producer gas, which is a very usable fuel.

Conversion of biomass to biofuel can also be achieved via selective conversion of individual components of biomass. For example, cellulose can be converted to intermediate platform chemical such a sorbitol, glucose, hydroxymethylfurfural etc. These chemical are then further reacted to produce hydrogen or hydrocarbon fuels.

Biomass also has the potential to be converted to multiple commodity chemicals. Halomethanes have successfully been by produced using a combination of A. fermentans and engineered S. cerevisiae. This method converts NaX salts and unprocessed biomass such as switchgrass, sugarcane, corn stover, or poplar into halomethanes. S-adenosylmethionine which is naturally occurring in S. cerevisiae allows a methyl group to be transferred. Production levels of 150 mg L-1H-1 iodomethane were achieved. At these levels roughly 173000L of capacity would need to be operated just to replace the United States' need for iodomethane. However, an advantage of this method is that it uses NaI rather than I2; NaI is significantly less hazardous than I2. This method may be applied to produce ethylene in the future.

Other chemical processes such as converting straight and waste vegetable oils into biodiesel is transesterification.

Biochemical Conversion

As biomass is a natural material, many highly efficient biochemical processes have developed in nature to break down the molecules of which biomass is composed, and many of these biochemical conversion processes can be harnessed.

Biochemical conversion makes use of the enzymes of bacteria and other microorganisms to break down biomass. In most cases, microorganisms are used to perform the conversion process: anaerobic digestion, fermentation, and composting.

Electrochemical Conversion

In addition to combustion, bio-mass/bio-fuels can be directly converted to electrical energy via electrochemical oxidation of the material. This can be performed directly in a direct carbon fuel cell, direct ethanol fuel cell or a microbial fuel cell. The fuel can also be consumed indirectly via a fuel cell system containing a reformer which converts the bio-mass into a mixture of CO and H_2 before it is consumed in the fuel cell.

In The United States

The biomass power generating industry in the United States, which consists of approximately 11,000 MW of summer operating capacity actively supplying power to the grid, produces about 1.4 percent of the U.S. electricity supply.

Currently, the New Hope Power Partnership is the largest biomass power plant in the U.S. The 140 MW facility uses sugarcane fiber (bagasse) and recycled urban wood as fuel to generate enough power for its large milling and refining operations as well as to supply electricity for nearly 60,000 homes.

Second-generation Biofuels

Second-generation biofuels were not (in 2010) produced commercially, but a considerable number of research activities were taking place mainly in North America, Europe and also in some emerging countries. These tend to use feedstock produced by rapidly reproducing enzymes or bacteria from various sources including excrement grown in Cell cultures or hydroponics There is huge potential for second generation biofuels but non-edible feedstock resources are highly under-utilized.

Environmental Impact

Using biomass as a fuel produces air pollution in the form of carbon monoxide, carbon dioxide, NOx (nitrogen oxides), VOCs (volatile organic compounds), particulates and other pollutants at levels above those from traditional fuel sources such as coal or natural gas in some cases (such as with indoor heating and cooking). Utilization of wood biomass as a fuel can also produce fewer particulate and other pollutants than open burning as seen in wildfires or direct heat applications. Black carbon – a pollutant created by combustion of fossil fuels, biofuels, and biomass – is possibly the second largest contributor to global warming. In 2009 a Swedish study of the giant brown haze that periodically covers large areas in South Asia determined that it had been principally produced by biomass burning, and to a lesser extent by fossil-fuel burning. Researchers measured a significant concentration of ^{14}C, which is associated with recent plant life rather than with fossil fuels.

Biomass power plant size is often driven by biomass availability in close proximity as transport costs of the (bulky) fuel play a key factor in the plant's economics. It has to be noted, however, that rail and especially shipping on waterways can reduce transport costs significantly, which has led to a global biomass market. To make small plants of 1 MW_{el} economically profitable those power plants need to be equipped with technology that is able to convert biomass to useful electricity with high efficiency such as ORC technology, a cycle similar to the water steam power process just with an organic working medium. Such small power plants can be found in Europe.

On combustion, the carbon from biomass is released into the atmosphere as carbon dioxide (CO_2). The amount of carbon stored in dry wood is approximately 50% by weight. However, according to the Food and Agriculture Organization of the United Nations, plant matter used as a fuel can be replaced by planting for new growth. When the biomass is from forests, the time to recapture the carbon stored is generally longer, and the carbon storage capacity of the forest may be reduced overall if destructive forestry techniques are employed.

Industry professionals claim that a range of issues can affect a plant's ability to comply with emissions standards. Some of these challenges, unique to biomass plants, include inconsistent fuel supplies and age. The type and amount of the fuel supply are completely reliant factors; the fuel can be in the form of building debris or agricultural waste (such as deforestation of invasive species or orchard trimmings). Furthermore, many of the biomass plants are old, use outdated technology and were not built to comply with today's stringent standards. In fact, many are based on technologies developed during the term of U.S. President Jimmy Carter, who created the United States Department of Energy in 1977.

The U.S. Energy Information Administration projected that by 2017, biomass is expected to be about twice as expensive as natural gas, slightly more expensive than nuclear power, and much less expensive than solar panels. In another EIA study released, concerning the government's plan to implement a 25% renewable energy standard by 2025, the agency assumed that 598 million tons of biomass would be available, accounting for 12% of the renewable energy in the plan.

The adoption of biomass-based energy plants has been a slow but steady process. Between the years of 2002 and 2012 the production of these plants has increased 14%. In the United States, alternative electricity-production sources on the whole generate about 13% of power; of this fraction, biomass contributes approximately 11% of the alternative production. According to a study conducted in early 2012, of the 107 operating biomass plants in the United States, 85 have been cited by federal or state regulators for the violation of clean air or water standards laws over the past 5 years. This data also includes minor infractions.

Despite harvesting, biomass crops may sequester carbon. For example, soil organic carbon has been observed to be greater in switchgrass stands than in cultivated cropland soil, especially at depths below 12 inches. The grass sequesters the carbon in its increased root biomass. Typically, perennial crops sequester much more carbon than annual crops due to much greater non-harvested living biomass, both living and dead, built up over years, and much less soil disruption in cultivation.

The proposal that biomass is carbon-neutral put forward in the early 1990s has been superseded by more recent science that recognizes that mature, intact forests sequester carbon more effectively than cut-over areas. When a tree's carbon is released into the atmosphere in a single pulse, it contributes to climate change much more than woodland timber rotting slowly over decades. Current studies indicate that "even after 50 years the forest has not recovered to its initial carbon storage" and "the optimal strategy is likely to be protection of the standing forest".

The pros and cons of biomass usage regarding carbon emissions may be quantified with the ILUC factor. There is controversy surrounding the usage of the ILUC factor.

Forest-based biomass has recently come under fire from a number of environmental organizations, including Greenpeace and the Natural Resources Defense Council, for the harmful impacts it can have on forests and the climate. Greenpeace recently released a report entitled "Fuelling a BioMess" which outlines their concerns around forest-based biomass. Because any part of the tree can be burned, the harvesting of trees for energy production encourages Whole-Tree Harvesting, which removes more nutrients and soil cover than regular harvesting, and can be harmful to the long-term health of the forest. In some jurisdictions, forest biomass removal is increasingly involving elements essential to functioning forest ecosystems, including standing trees, naturally disturbed forests and remains of traditional logging operations that were previously left in the forest. Environmental groups also cite recent scientific research which has found that it can take many decades for the carbon released by burning biomass to be recaptured by regrowing trees, and even longer in low productivity areas; furthermore, logging operations may disturb forest soils and cause them to release stored carbon. In light of the pressing need to reduce greenhouse gas emissions in the short term in order to mitigate the effects of climate change, a number of environmental groups are opposing the large-scale use of forest biomass in energy production.

Supply Chain Issues

With the seasonality of biomass supply and a great variability in sources, supply chains play a key role in cost-effective delivery of bioenergy. There are several potential challenges peculiar to bioenergy supply chains:

Technical issues

- Inefficiencies of conversion

- Storage methods for seasonal availability

- Complex multi-component constituents incompatible with maximizing efficiency of single purpose use

- High water content

- Conflicting decisions (technologies, locations, and routes)

- Complex location analysis (source points, inventory facilities, and production plants)

Logistic issues

- Seasonal availability leading to storage challenges and/or seasonally idle facilities

- Low bulk-density and/or high water content

- Finite productivity per area and/or time incompatible with conventional approach to economy of scale focusing on maximizing facility size

Financial issues

- The limits for the traditional approach to economy of scale which focuses on maximizing single facility size

- Unavailability and complexity of life cycle costing data

- Lack of required transport infrastructure

- Limited flexibility or inflexibility to energy demand

- Risks associated with new technologies (insurability, performance, rate of return)

- Extended market volatilities (conflicts with alternative markets for biomass)

- Difficult or impossible to use financial hedging methods to control cost

Social issues

- Lack of participatory decision making

- Lack of public/community awareness

- Local supply chain impacts vs. global benefits

- Health and safety risks

- Extra pressure on transport sector

- Decreasing the esthetics of rural areas

Policy and regulatory issues

- Impact of fossil fuel tax on biomass transport

- Lack of incentives to create competition among bioenergy producers

- Focus on technology options and less attention to selection of biomass materials

- Lack of support for sustainable supply chain solutions

Institutional and organizational issues

- Varied ownership arrangements and priorities among supply chain parties

- Lack of supply chain standards

- Impact of organizational norms and rules on decision making and supply chain coordination

- Immaturity of change management practices in biomass supply chains

Renewable Natural Gas

Renewable natural gas, also known as sustainable natural gas, is a biogas which has been upgraded to a quality similar to fossil natural gas. A biogas is a gas methane obtained from biomass. By upgrading the quality to that of natural gas, it becomes possible to distribute the gas to customers via the existing gas grid, within existing appliances. Renewable natural gas is a subset of synthetic natural gas or substitute natural gas (SNG).

Renewable natural gas can be produced economically, and distributed via the existing gas grid, making it an attractive means of supplying existing premises with renewable heat and renewable gas energy, while requiring no extra capital outlay of the customer. Renewable natural gas can be converted into liquefied natural gas (LNG) for direct use as fuel in transport sector. LNG would fetch good price equivalent to gasoline or diesel as it can replace these fuels in transport sector.

The existing gas network allows distribution of gas energy over vast distances at a minimal cost in energy. Existing networks would allow biogas to be sourced from remote markets that are rich in low-cost biomass (Russia or Scandinavia for example).

The UK National Grid believes that at least 15% of all gas consumed could be made from matter such as sewage, food waste such as food thrown away by supermarkets and restaurants, and organic waste created by businesses such as breweries.

Manufacturing

A biomass to SNG efficiency of 70% can be achieved. Costs are minimised by maximising production scale, and by locating plant next to transport links (e.g. a port or river) for the chosen source of biomass. The existing gas storage infrastructure would allow the plant to continue to manufacture gas at the full utilisation rate even during periods of weak demand, helping minimise manufacturing capital costs per unit of gas produced.

Renewable gas can be produced through three main processes; anaerobic digestion of organic (normally moist) material, thermal gasification of organic (normally dry) material and produced through the Sabatier reaction. In these cases the gas from primary production has to be upgraded in a secondary step to produce gas that is suitable for injection into the gas grid.

Commercial Development

BioSNG

Göteborg Energi successfully opened the first demonstration plant for large scale production of bio-SNG through gasification of forest residues in Gothenburg, Sweden within the GoBiGas project. The plant has a capacity to produce 20 megawatts-worth of bioSNG from about 30 MW-worth of biomass, aiming at a conversion efficiency of 65%. Since December 2014 the bioSNG plant is fully operational and gas is supplied to the Swedish natural gas grid, reaching the quality demands with a methane content of over 95%.

SNG is of particular interest in countries with extensive natural gas distribution networks. Core advantages of SNG include compatibility with existing natural gas infrastructure, higher efficiency that Fisher-Tropsch fuels production and smaller-production scale than other second generation biofuel production systems. The Energy Research Centre of the Netherlands has conducted extensive research on large-scale SNG production from woody biomass, based on the importation of feedstocks from abroad.

Renewable natural gas plants based on wood can be categorised into two main categories, one being allothermal, which has the energy provided by a source outside of the gasifier. One example is the double-chambered fluidised bed gasifiers consisting of a separate combustion and gasification chambers. Autothermal systems generate the heat within the gasifier, but require the use of pure oxygen to avoid nitrogen dilution.

In the UK, the National Non-Food Crops Centre found that any UK bioSNG plant built by 2020 would be highly likely to use 'clean woody feedstocks' and that there are several regions with good availability of that source.

Upgraded Biogas

In the UK, using anaerobic digestion is growing as a means of producing renewable biogas, with nearly 50 sites built across the country. Ecotricity has announced plans to supply green gas to UK consumers via the national grid. Centrica has also announced that it will soon begin injecting gas, manufactured from sewage, into the gas grid. In Canada, FortisBC, a gas provider in British Columbia, has begun injecting limited amounts of renewably created natural gas into its existing gas distribution system to begin to offer customers renewable gas options.

Sustainable Synthetic Natural Gas

Sustainable SNG is produced by high temperature oxygen blown slagging co-gasification at 70 to 75 bar pressure of liquid and solid contaminated and wood, biomass, negative cost hazardous and non-hazardous wastes, coal and natural gas. This uses coal to SNG technology developed from the end of WW2 onwards, and successfully demonstrated at SVZ Schwarze Pumpe. The same technology can be transferred from the low grade lignite to fertiliser industry, where it is currently being successfully developed in China, to the renewable energy industry.

The advantage of a wide range of feedstocks is that much larger quantities of renewable SNG can be produced compared with biogas, with fewer supply chain limitations. A wide range of fuels with an overall biogenic carbon content of 50 to 55% is technically and financially viable. Hydrogen is added to the fuel mix during the gasification process, and carbon dioxide is removed by capture from the purge gas 'slip stream' syngas clean-up and catalytic methanation stages.

Large scale sustainable SNG will enable the UK gas and electricity grids to be substantially de-carbonised in parallel at source, while maintaining the existing operational and economic relationship between the gas and electricity grids. Carbon capture and sequestration can be added at little additional cost, thereby progressively achieving deeper de-carbonisation of the existing gas and electricity grids at low cost and operational risk. Cost benefit studies indicate that large scale 50% biogenic carbon content sustainable SNG can be injected into the high pressure gas transmission grid at a cost of around 65p/therm. At this cost, it is possible to re-process fossil natural gas, used as an energy input into the gasification process, into 5 to 10 times greater quantity of sustainable SNG. Large scale sustainable SNG, combined with continuing natural gas production from UK continental shelf and unconventional gas, will potentially enable the cost of UK peak electricity to be de-coupled from international oil denominated 'take or pay' gas supply contracts.

Applications:

- Electricity generation

- Space heating

- Process heating

- Biomass with carbon capture and storage

- Transportation fuel

Biochar

Biochar is charcoal used as a soil amendment. Like most charcoal, biochar is made from biomass via pyrolysis. Biochar is under investigation as an approach to carbon sequestration to produce negative carbon dioxide emissions. Biochar thus has the potential to help mitigate climate change via carbon sequestration. Independently, biochar can increase soil fertility of acidic soils (low pH soils), increase agricultural productivity, and provide protection against some foliar and soil-borne diseases. Furthermore, biochar reduces pressure on forests. Biochar is a stable solid, rich in carbon, and can endure in soil for thousands of years.

A piece of biochar.

History

Pre-Columbian Amazonians are believed to have used biochar to enhance soil productivity. They produced it by smoldering agricultural waste (i.e., covering burning biomass with soil) in pits or trenches. European settlers called it *terra preta de Indio*. Following observations and experiments, a research team working in French Guiana hypothesized that the Amazonian earthworm *Pontoscolex corethrurus* was the main agent of fine powdering and incorporation of charcoal debris to the mineral soil.

The term "biochar" was coined by Peter Read to describe charcoal used as a soil improvement.

Production

Biochar is a high-carbon, fine-grained residue that today is produced through modern pyrolysis processes, which is the direct thermal decomposition of biomass in the absence of oxygen, which prevents combustion, to obtain an array of solid (biochar), liquid (bio-oil), and gas (syngas) products. The specific yield from the pyrolysis is dependent on process conditions. such as temperature, and can be optimized to produce either energy or biochar. Temperatures of 400–500 °C (752–932 °F) produce more char, while temperatures above 700 °C (1,292 °F) favor the yield of liquid and gas fuel components. Pyrolysis occurs more quickly at the higher temperatures, typically requiring seconds instead of hours. High temperature pyrolysis is also known as gasification, and produces primarily syngas. Typical yields are 60% bio-oil, 20% biochar, and 20% syngas. By comparison, slow pyrolysis can produce substantially more char (~50%). Once initialized, both processes produce net energy. For typical inputs, the energy required to run a "fast" pyrolyzer is approximately 15% of the energy that it outputs. Modern pyrolysis plants can use the syngas created by the pyrolysis process and output 3–9 times the amount of energy required to run.

The Amazonian pit/trench method harvests neither bio-oil nor syngas, and releases a large amount of CO_2, black carbon, and other greenhouse gases (GHG)s (and potentially, toxins) into the air. Commercial-scale systems process agricultural waste, paper byproducts, and even municipal waste and typically eliminate these side effects by capturing and using the liquid and gas products.

Centralized, Decentralized, and Mobile Systems

In a centralized system, all biomass in a region is brought to a central plant for processing. Alternatively, each farmer or group of farmers can operate a lower-tech kiln. Finally, a truck equipped with a pyrolyzer can move from place to place to pyrolyze biomass. Vehicle power comes from the syngas stream, while the biochar remains on the farm. The biofuel is sent to a refinery or storage site. Factors that influence the choice of system type include the cost of transportation of the liquid and solid byproducts, the amount of material to be processed, and the ability to feed directly into the power grid.

For crops that are not exclusively for biochar production, the residue-to-product ratio (RPR) and the collection factor (CF) the percent of the residue not used for other things, measure the approximate amount of feedstock that can be obtained for pyrolysis after harvesting the primary product. For instance, Brazil harvests approximately 460 million tons (MT) of sugarcane annually, with an RPR of 0.30, and a CF of 0.70 for the sugarcane tops, which normally are burned in the field. This translates into approximately 100 MT of residue annually, which could be pyrolyzed to create energy and soil additives. Adding in the bagasse (sugarcane waste) (RPR=0.29 CF=1.0), which is otherwise burned (inefficiently) in boilers, raises the total to 230 MT of pyrolysis feedstock. Some plant residue, however, must remain on the soil to avoid increased costs and emissions from nitrogen fertilizers.

Pyrolysis technologies for processing loose and leafy biomass produce both biochar and syngas.

Thermo-catalytic Depolymerization

Alternatively, "thermo-catalytic depolymerization", which utilizes microwaves, has recently been used to efficiently convert organic matter to biochar on an industrial scale, producing ~50% char.

Uses

Carbon Sink

The burning and natural decomposition of biomass and in particular agricultural waste adds large amounts of CO2 to the atmosphere. Biochar that is stable, fixed, and 'recalcitrant' carbon can store large amounts of greenhouse gases in the ground for centuries, potentially reducing or stalling the growth in atmospheric greenhouse gas levels; at the same time its presence in the earth can improve water quality, increase soil fertility, raise agricultural productivity, and reduce pressure on old-growth forests.

Biochar can sequester carbon in the soil for hundreds to thousands of years, like coal. Such a carbon-negative technology would lead to a net withdrawal of CO_2 from the atmosphere, while producing and consuming energy. This technique is advocated by prominent scientists such as James Hansen, head of the NASA Goddard Institute for Space Studies, and James Lovelock, creator of the Gaia hypothesis, for mitigation of global warming by greenhouse gas remediation.

Researchers have estimated that sustainable use of biocharring could reduce the global net emissions of carbon dioxide (CO2), methane, and nitrous oxide by up to 1.8 Pg

CO2-C equivalent (CO2-C$_e$) per year (12% of current anthropogenic CO2-Ce emissions; 1 Pg=1 Gt), and total net emissions over the course of the next century by 130 Pg CO2-C$_e$, without endangering food security, habitat, or soil conservation.

Soil Amendment

Biochar is recognised as offering a number of benefits for soil health. Many benefits are related to the extremely porous nature of biochar. This structure is found to be very effective at retaining both water and water-soluble nutrients. Soil biologist Elaine Ingham indicates the extreme suitability of biochar as a habitat for many beneficial soil micro organisms. She points out that when pre charged with these beneficial organisms biochar becomes an extremely effective soil amendment promoting good soil, and in turn plant, health.

Biochar has also been shown to reduce leaching of *E-coli* through sandy soils depending on application rate, feedstock, pyrolysis temperature, soil moisture content, soil texture, and surface properties of the bacteria.

For plants that require high potash and elevated pH, biochar can be used as a soil amendment to improve yield.

Biochar can improve water quality, reduce soil emissions of greenhouse gases, reduce nutrient leaching, reduce soil acidity, and reduce irrigation and fertilizer requirements. Biochar was also found under certain circumstances to induce plant systemic responses to foliar fungal diseases and to improve plant responses to diseases caused by soilborne pathogens.

The various impacts of biochar can be dependent on the properties of the biochar, as well as the amount applied, and there is still a lack of knowledge about the important mechanisms and properties. Biochar impact may depend on regional conditions including soil type, soil condition (depleted or healthy), temperature, and humidity. Modest additions of biochar to soil reduce nitrous oxide N2O emissions by up to 80% and eliminate methane emissions, which are both more potent greenhouse gases than CO_2.

Studies have reported positive effects from biochar on crop production in degraded and nutrient–poor soils. Biochar can be designed with specific qualities to target distinct properties of soils. Biochar reduces leaching of critical nutrients, creates a higher crop uptake of nutrients, and provides greater soil availability of nutrients. At 10% levels biochar reduced contaminant levels in plants by up to 80%, while reducing total chlordane and DDX content in the plants by 68 and 79%, respectively. On the other hand, because of its high adsorption capacity, biochar may reduce the efficacy of soil applied pesticides that are needed for weed and pest control. High-surface-area biochars may be particularly problematic in this regard; more research into the long-term effects of biochar addition to soil is needed.

Slash-and-char

Switching from *slash-and-burn* to *slash-and-char* farming techniques in Brazil can decrease both deforestation of the Amazon basin and carbon dioxide emission, as well as increase crop yields. Slash-and-burn leaves only 3% of the carbon from the organic material in the soil.

Slash-and-char can keep up to 50% of the carbon in a highly stable form. Returning the biochar into the soil rather than removing it all for energy production reduces the need for nitrogen fertilizers, thereby reducing cost and emissions from fertilizer production and transport. Additionally, by improving the soil's ability to be tilled, fertility, and productivity, biochar–enhanced soils can indefinitely sustain agricultural production, whereas non-enriched soils quickly become depleted of nutrients, forcing farmers to abandon the fields, producing a continuous slash and burn cycle and the continued loss of tropical rainforest. Using pyrolysis to produce bio-energy also has the added benefit of not requiring infrastructure changes the way processing biomass for cellulosic ethanol does. Additionally, the biochar produced can be applied by the currently used machinery for tilling the soil or equipment used to apply fertilizer.

Water Retention

Biochar is a desirable soil material in many locations due to its ability to attract and retain water. This is possible because of its porous structure and high surface area. As a result, nutrients, phosphorus, and agrochemicals are retained for the plants benefit. Plants therefore, are healthier and fertilizers leach less into surface or groundwater.

Energy Production: Bio-oil and Syngas

Mobile pyrolysis units can be used to lower the costs of transportation of the biomass if the biochar is returned to the soil and the syngas stream is used to power the process. Bio-oil contains organic acids that are corrosive to steel containers, has a high water vapor content that is detrimental to ignition, and, unless carefully cleaned, contains some biochar particles which can block injectors.

If biochar is used for the production of energy rather than as a soil amendment, it can be directly substituted for any application that uses coal. Pyrolysis also may be the most cost-effective way of electricity generation from biomaterial.

Direct and Indirect Benefits

- The pyrolysis of forest- or agriculture-derived biomass residue generates a bio-fuel without competition with crop production.

- Biochar is a pyrolysis byproduct that may be ploughed into soils in crop fields to enhance their fertility and stability, and for medium- to long-term carbon sequestration in these soils.

- Biochar enhances the natural process: the biosphere captures CO_2, especially through plant production, but only a small portion is stably sequestered for a relatively long time (soil, wood, etc.).

- Biomass production to obtain biofuels and biochar for carbon sequestration in the soil is a carbon-negative process, i.e. more CO_2 is removed from the atmosphere than released, thus enabling long-term sequestration.

Research

Intensive research into manifold aspects involving the pyrolysis/biochar platform is underway around the world. From 2005 to 2012, there were 1,038 articles that included the word "biochar" or "bio-char" in the topic that had been indexed in the ISI Web of Science. Further research is in progress by such diverse institutions around the world as Cornell University, the University of Edinburgh, which has a dedicated research unit., and the Agricultural Research Organization (ARO) of Israel, Volcani Center, where a network of researchers involved in biochar research (iBRN, Israel Biochar Researchers Network) was established as early as 2009.

Students at Stevens Institute of Technology in New Jersey are developing supercapacitors that use electrodes made of biochar. A process developed by University of Florida researchers that removes phosphate from water, also yields methane gas usable as fuel and phosphate-laden carbon suitable for enriching soil.

Emerging Commercial Sector

Calculations suggest that emissions reductions can be 12 to 84% greater if biochar is put back into the soil instead of being burned to offset fossil-fuel use. Thus biochar sequestration offers the chance to turn bioenergy into a carbon-negative industry.

Johannes Lehmann, of Cornell University, estimates that pyrolysis can be cost-effective for a combination of sequestration and energy production when the cost of a CO_2 ton reaches $37. As of mid-February 2010, CO_2 is trading at $16.82/ton on the European Climate Exchange (ECX), so using pyrolysis for bioenergy production may be feasible even if it is more expensive than fossil fuel.

Current biochar projects make no significant impact on the overall global carbon budget, although expansion of this technique has been advocated as a geoengineering approach. In May 2009, the Biochar Fund received a grant from the Congo Basin Forest Fund for a project in Central Africa to simultaneously slow down deforestation, increase the food security of rural communities, provide renewable energy and sequester carbon.

Application rates of 2.5–20 tonnes per hectare (1.0–8.1 t/acre) appear to be required to produce significant improvements in plant yields. Biochar costs in developed countries vary from \$300–7000/tonne, generally too high for the farmer/horticulturalist and prohibitive for low-input field crops. In developing countries, constraints on agricultural biochar relate more to biomass availability and production time. An alternative is to use small amounts of biochar in lower cost biochar-fertilizer complexes.

Various companies in North America, Australia, and England sell biochar or biochar production units. In England Carbon Gold supply a range of biochar-based soil improvers, composts and fertilisers for arboriculture, horticulture and turfcare as well as to home growers. In Sweden the 'Stockholm Solution' is an urban tree planting system that uses 30% biochar to support healthy growth of the urban forest. The Qatar Aspire Park now uses biochar to help trees cope with the intense heat of their summers.

At the 2009 International Biochar Conference, a mobile pyrolysis unit with a specified intake of 1,000 pounds (450 kg) was introduced for agricultural applications. The unit had a length of 12 feet and height of 7 feet (3.6 m by 2.1m).

A production unit in Dunlap, Tennessee by Mantria Corporation opened in August 2009 after testing and an initial run, was later shut down as part of a Ponzi scheme investigation.

References

- Starke, Linda (2009). State Of The World 2009: Into a Warming World: a WorldWatch Institute Report on Progress Toward a Sustainable Society. WW Norton & Company. ISBN 978-0-393-33418-0.

- "Learning About Renewable Energy". NREL's vision is to develop technology. National Renewable Energy Laboratory. Retrieved 4 April 2013.

- Biederman, Lori A.; W. Stanley Harpole (2011). "Biochar and Managed Perennial Ecosystems". Iowa State Research Farm Progress Reports. Retrieved February 12, 2013.

- Brewer, Catherine (2012). Biochar Characterization and Engineering (dissertation). Iowa State University. Retrieved February 12, 2013.

- Forest volume-to-biomass models and estimates of mass for live and standing dead trees of U.S. forests. (PDF) . Retrieved on 2012-02-28.

- Scheck, Justin; et al. (July 23, 2012). "Wood-Fired Plants Generate Violations". Wall Street Journal. Retrieved September 27, 2012.

- Van der Meijden, C.M. (2010). Development of the MILENA gasification technology for the production of Bio-SNG (PDF). Petten, Netherlands: ECN. Retrieved 21 October 2012.

4

Fuels: Sources of Bioenergy

This chapter focuses on the sources of bioenergy. The fuels explained in this section are wood, straw, manure and sugarcane. Wood has been used as a fuel for thousands of years whereas straw is an agricultural by-product and makes up for about half of the yield of cereal crops. The topics discussed in this chapter are of vital importance and provides a better understanding of bioenergy.

Wood

Wood is a porous and fibrous structural tissue found in the stems and roots of trees, and other woody plants. It has been used for thousands of years for both fuel and as a construction material. It is an organic material, a natural composite of cellulose fibers (which are strong in tension) embedded in a matrix of lignin which resists compression. Wood is sometimes defined as only the secondary xylem in the stems of trees, or it is defined more broadly to include the same type of tissue elsewhere such as in the roots of trees or shrubs. In a living tree it performs a support function, enabling woody plants to grow large or to stand up by themselves. It also conveys water and nutrients between the leaves, other growing tissues, and the roots. Wood may also refer to other plant materials with comparable properties, and to material engineered from wood, or wood chips or fiber.

In 2005, the Earth contained about 434 billion cubic meters of growing stock forest, 47% of which was commercial. As an abundant, carbon-neutral renewable resource, woody materials have been of intense interest as a source of renewable energy. In 1991 approximately 3.5 cubic kilometers of wood were harvested. Dominant uses were for furniture and building construction.

History

A 2011 discovery in the Canadian province of New Brunswick uncovered the earliest known plants to have grown wood, approximately 395 to 400 million years ago. Wood can be dated by carbon dating and in some species by dendrochronology to make inferences about when a wooden object was created.

People have used wood for millennia for many purposes, primarily as a fuel or as a construction material for making houses, tools, weapons, furniture, packaging, artworks, and paper. The year-to-year variation in tree-ring widths and isotopic abundances gives clues to the prevailing climate at that time.

Physical Properties

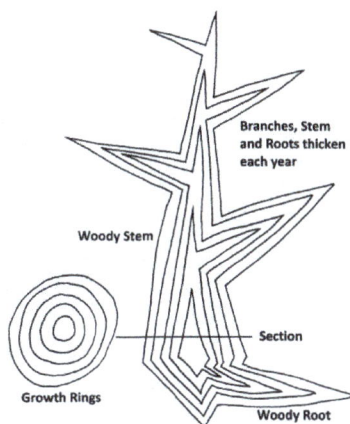

Diagram of secondary growth in a tree showing idealised vertical and horizontal sections. A new layer of wood is added in each growing season, thickening the stem, existing branches and roots, to form a growth ring.

Growth Rings

Wood, in the strict sense, is yielded by trees, which increase in diameter by the formation, between the existing wood and the inner bark, of new woody layers which envelop the entire stem, living branches, and roots. This process is known as secondary growth; it is the result of cell division in the vascular cambium, a lateral meristem, and subsequent expansion of the new cells. These cells then go on to form thickened secondary cell walls, composed mainly of cellulose, hemicellulose and lignin. Where the differences between the four seasons are distinct, growth can occur in a discrete annual or seasonal pattern, leading to growth rings; these can usually be most clearly seen on the end of a log, but are also visible on the other surfaces. If the distinctiveness between seasons is annual, these growth rings are referred to as annual rings. Where there is little seasonal difference growth rings are likely to be indistinct or absent.

If there are differences within a growth ring, then the part of a growth ring nearest the center of the tree, and formed early in the growing season when growth is rapid, is usually composed of wider elements. It is usually lighter in color than that near the outer portion of the ring, and is known as earlywood or springwood. The outer portion formed later in the season is then known as the latewood or summerwood. However, there are major differences, depending on the kind of wood.

Knots

A knot is a particular type of imperfection in a piece of wood; it will affect the technical properties of the wood, usually reducing the local strength and increasing the tendency for splitting along the wood grain, but may be exploited for visual effect. In a longitudinally sawn plank, a knot will appear as a roughly circular "solid" (usually darker) piece of wood around which the grain of the rest of the wood "flows" (parts and rejoins). Within a knot, the direction of the wood (grain direction) is up to 90 degrees different from the grain direction of the regular wood.

A knot on a tree at the Garden of the Gods public park in Colorado Springs, Colorado (October 2006)

In the tree a knot is either the base of a side branch or a dormant bud. A knot (when the base of a side branch) is conical in shape (hence the roughly circular cross-section) with the inner tip at the point in stem diameter at which the plant's vascular cambium was located when the branch formed as a bud.

During the development of a tree, the lower limbs often die, but may remain attached for a time, sometimes years. Subsequent layers of growth of the attaching stem are no longer intimately joined with the dead limb, but are grown around it. Hence, dead branches produce knots which are not attached, and likely to drop out after the tree has been sawn into boards.

In grading lumber and structural timber, knots are classified according to their form, size, soundness, and the firmness with which they are held in place. This firmness is affected by, among other factors, the length of time for which the branch was dead while the attaching stem continued to grow.

Wood knot in vertical section

Knots materially affect cracking and warping, ease in working, and cleavability of timber. They are defects which weaken timber and lower its value for structural purposes where strength is an important consideration. The weakening effect is much more serious when timber is subjected to forces perpendicular to the grain and/or tension than where under load along the grain and/or compression. The extent to which knots affect the strength of a beam depends upon their position, size, number, and condition. A knot on the upper side is compressed, while one on the lower side is subjected to tension. If there is a season check in the knot, as is often the case, it will offer little resistance to this tensile stress. Small knots, however, may be located along the neutral plane of a beam and increase the strength by preventing longitudinal shearing. Knots in a board or plank are least injurious when they extend through it at right angles to its broadest surface. Knots which occur near the ends of a beam do not weaken it. Sound knots which occur in the central portion one-fourth the height of the beam from either edge are not serious defects.

—Samuel J. Record, The Mechanical Properties of Wood

Knots do not necessarily influence the stiffness of structural timber, this will depend on the size and location. Stiffness and elastic strength are more dependent upon the sound wood than upon localized defects. The breaking strength is very susceptible to defects. Sound knots do not weaken wood when subject to compression parallel to the grain.

In some decorative applications, wood with knots may be desirable to add visual interest. In applications where wood is painted, such as skirting boards, fascia boards, door frames and furniture, resins present in the timber may continue to 'bleed' through to the surface of a knot for months or even years after manufacture and show as a yellow or brownish stain. A knot primer paint or solution (knotting), correctly applied during preparation, may do much to reduce this problem but it is difficult to control complete-ly, especially when using mass-produced kiln-dried timber stocks.

Heartwood and Sapwood

Heartwood (or duramen) is wood that as a result of a naturally occurring chemical transformation has become more resistant to decay. Heartwood formation occurs spontaneously (it is a genetically programmed process). Once heartwood formation is complete, the heartwood is dead. Some uncertainty still exists as to whether heartwood is truly dead, as it can still chemically react to decay organisms, but only once.

A section of a Yew branch showing 27 annual growth rings, pale sapwood, dark heartwood, and pith (center dark spot). The dark radial lines are small knots.

Heartwood is often visually distinct from the living sapwood, and can be distinguished in a cross-section where the boundary will tend to follow the growth rings. For exam-ple, it is sometimes much darker. However, other processes such as decay or insect in-vasion can also discolor wood, even in woody plants that do not form heartwood, which may lead to confusion.

Sapwood (or alburnum) is the younger, outermost wood; in the growing tree it is living wood, and its principal functions are to conduct water from the roots to the leaves and to store up and give back according to the season the reserves prepared in the leaves. However, by the time they become competent to conduct water, all xylem tracheids and vessels have lost their cytoplasm and the cells are therefore functionally dead. All wood in a tree is first formed as sapwood. The more leaves a tree bears and the more vigor-ous its growth, the larger the volume of sapwood required. Hence trees making rapid growth in the open have thicker sapwood for their size than trees of the same species growing in dense forests. Sometimes trees (of species that do form heartwood) grown in the open may become of considerable size, 30 cm (12 in) or more in diameter, before any heartwood begins to form, for example, in second-growth hickory, or open-grown pines.

The term *heartwood* derives solely from its position and not from any vital importance to the tree. This is evidenced by the fact that a tree can thrive with its heart completely decayed. Some species begin to form heartwood very early in life, so having only a thin layer of live sapwood, while in others the change comes slowly. Thin sapwood is characteristic of such species as chestnut, black locust, mulberry, osage-orange, and sassafras, while in maple, ash, hickory, hackberry, beech, and pine, thick sapwood is the rule. Others never form heartwood.

No definite relation exists between the annual rings of growth and the amount of sapwood. Within the same species the cross-sectional area of the sapwood is very roughly proportional to the size of the crown of the tree. If the rings are narrow, more of them are required than where they are wide. As the tree gets larger, the sapwood must necessarily become thinner or increase materially in volume. Sapwood is thicker in the upper portion of the trunk of a tree than near the base, because the age and the diameter of the upper sections are less.

When a tree is very young it is covered with limbs almost, if not entirely, to the ground, but as it grows older some or all of them will eventually die and are either broken off or fall off. Subsequent growth of wood may completely conceal the stubs which will however remain as knots. No matter how smooth and clear a log is on the outside, it is more or less knotty near the middle. Consequently, the sapwood of an old tree, and particularly of a forest-grown tree, will be freer from knots than the inner heartwood. Since in most uses of wood, knots are defects that weaken the timber and interfere with its ease of working and other properties, it follows that a given piece of sapwood, because of its position in the tree, may well be stronger than a piece of heartwood from the same tree.

It is remarkable that the inner heartwood of old trees remains as sound as it usually does, since in many cases it is hundreds, and in a few instances thousands, of years old. Every broken limb or root, or deep wound from fire, insects, or falling timber, may afford an entrance for decay, which, once started, may penetrate to all parts of the trunk. The larvae of many insects bore into the trees and their tunnels remain indefinitely as sources of weakness. Whatever advantages, however, that sapwood may have in this connection are due solely to its relative age and position.

If a tree grows all its life in the open and the conditions of soil and site remain unchanged, it will make its most rapid growth in youth, and gradually decline. The annual rings of growth are for many years quite wide, but later they become narrower and narrower. Since each succeeding ring is laid down on the outside of the wood previously formed, it follows that unless a tree materially increases its production of wood from year to year, the rings must necessarily become thinner as the trunk gets wider. As a tree reaches maturity its crown becomes more open and the annual wood production is lessened, thereby reducing still more the width of the growth rings. In the case of forest-grown trees so much depends upon the competition of the trees in their struggle for light and nourishment that periods of rapid and slow growth may alternate. Some trees, such as southern oaks, maintain the same width of ring for hundreds of years. Upon the whole, however, as a tree gets larger in diameter the width of the growth rings decreases.

Different pieces of wood cut from a large tree may differ decidedly, particularly if the tree is big and mature. In some trees, the wood laid on late in the life of a tree is softer, lighter, weaker, and more even-textured than that produced earlier, but in other trees, the reverse applies. This may or may not correspond to heartwood and sapwood. In a large log the sapwood, because of the time in the life of the tree when it was grown, may be inferior in hardness, strength, and toughness to equally sound heartwood from the same log. In a smaller tree, the reverse may be true.

Color

In species which show a distinct difference between heartwood and sapwood the natural color of heartwood is usually darker than that of the sapwood, and very frequently the contrast is conspicuous. This is produced by deposits in the heartwood of chemical substances, so that a dramatic color difference does not mean a dramatic difference in the mechanical properties of heartwood and sapwood, although there may be a dramatic chemical difference.

The wood of Coast Redwood is distinctively red.

Some experiments on very resinous Longleaf Pine specimens indicate an increase in strength, due to the resin which increases the strength when dry. Such resin-saturated heartwood is called "fat lighter". Structures built of fat lighter are almost impervious to rot and termites; however they are very flammable. Stumps of old longleaf pines are often dug, split into small pieces and sold as kindling for fires. Stumps thus dug may actually remain a century or more since being cut. Spruce impregnated with crude resin and dried is also greatly increased in strength thereby.

Since the latewood of a growth ring is usually darker in color than the earlywood, this fact may be used in judging the density, and therefore the hardness and strength of the material. This is particularly the case with coniferous woods. In ring-porous woods the vessels of the early wood not infrequently appear on a finished surface as darker than the denser latewood, though on cross sections of heartwood the reverse is commonly true. Except in the manner just stated the color of wood is no indication of strength.

Abnormal discoloration of wood often denotes a diseased condition, indicating unsoundness. The black check in western hemlock is the result of insect attacks. The reddish-brown streaks so common in hickory and certain other woods are mostly the result of injury by birds. The discoloration is merely an indication of an injury, and in all probability does not of itself affect the properties of the wood. Certain rot-producing fungi impart to wood characteristic colors which thus become symptomatic of weakness; however an attractive effect known as spalting produced by this process is often considered a desirable characteristic. Ordinary sap-staining is due to fungal growth, but does not necessarily produce a weakening effect.

Water Content

Water occurs in living wood in three conditions, namely:

1. in the cell walls,

2. in the protoplasmic contents of the cells, and

3. as free water in the cell cavities and spaces.

In heartwood it occurs only in the first and last forms. Wood that is thoroughly air-dried retains 8–16% of the water in the cell walls, and none, or practically none, in the other forms. Even oven-dried wood retains a small percentage of moisture, but for all except chemical purposes, may be considered absolutely dry.

The general effect of the water content upon the wood substance is to render it softer and more pliable. A similar effect of common observation is in the softening action of water on rawhide, paper, or cloth. Within certain limits, the greater the water content, the greater its softening effect.

Drying produces a decided increase in the strength of wood, particularly in small specimens. An extreme example is the case of a completely dry spruce block 5 cm in section, which will sustain a permanent load four times as great as a green (undried) block of the same size will.

The greatest strength increase due to drying is in the ultimate crushing strength, and strength at elastic limit in endwise compression; these are followed by the modulus of rupture, and stress at elastic limit in cross-bending, while the modulus of elasticity is least affected.

Structure

Wood is a heterogeneous, hygroscopic, cellular and anisotropic material. It consists of cells, and the cell walls are composed of micro-fibrils of cellulose (40% – 50%) and hemicellulose (15% – 25%) impregnated with lignin (15% – 30%).

In coniferous or softwood species the wood cells are mostly of one kind, tracheids, and as a result the material is much more uniform in structure than that of most hardwoods. There are no vessels ("pores") in coniferous wood such as one sees so prominently in oak and ash, for example.

Sections of tree trunk

The structure of hardwoods is more complex. The water conducting capability is mostly taken care of by vessels: in some cases (oak, chestnut, ash) these are quite large and distinct, in others (buckeye, poplar, willow) too small to be seen without a hand lens. In discussing such woods it is customary to divide them into two large classes, *ring-porous* and *diffuse-porous*.

In ring-porous species, such as ash, black locust, catalpa, chestnut, elm, hickory, mulberry, and oak, the larger vessels or pores (as cross sections of vessels are called) are localised in the part of the growth ring formed in spring, thus forming a region of more or less open and porous tissue. The rest of the ring, produced in summer, is made up of smaller vessels and a much greater proportion of wood fibers. These fibers are the elements which give strength and toughness to wood, while the vessels are a source of weakness.

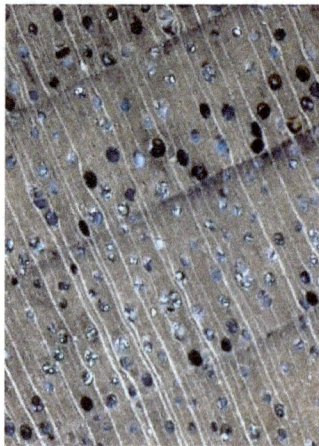

Magnified cross-section of Black Walnut, showing the vessels, rays (white lines) and annual rings: this is intermediate between diffuse-porous and ring-porous, with vessel size declining gradually

In diffuse-porous woods the pores are evenly sized so that the water conducting capability is scattered throughout the growth ring instead of being collected in a band or row. Examples of this kind of wood are alder, basswood, birch, buckeye, maple, willow,and the Populus species such as aspen, cottonwood and poplar. Some species, such as walnut and cherry, are on the border between the two classes, forming an intermediate group.

Earlywood and Latewood

In Softwood

In temperate softwoods there often is a marked difference between latewood and earlywood. The latewood will be denser than that formed early in the season. When examined under a microscope the cells of dense latewood are seen to be very thick-walled and with very small cell cavities, while those formed first in the season have thin walls and large cell cavities. The strength is in the walls, not the cavities. Hence the greater the proportion of latewood the greater the density and strength. In choosing a piece of pine where strength or stiffness is the important consideration, the principal thing to observe is the comparative amounts of earlywood and latewood. The width of ring is not nearly so important as the proportion and nature of the latewood in the ring.

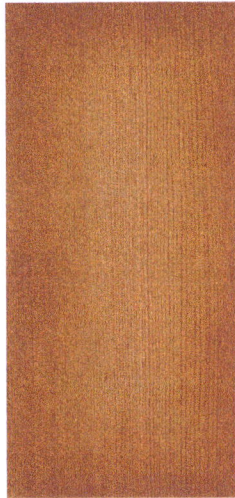

Earlywood and latewood in a softwood; radial view, growth rings closely spaced in Rocky Mountain Douglas-fir

If a heavy piece of pine is compared with a lightweight piece it will be seen at once that the heavier one contains a larger proportion of latewood than the other, and is therefore showing more clearly demarcated growth rings. In white pines there is not much contrast between the different parts of the ring, and as a result the wood is very uniform in texture and is easy to work. In hard pines, on the other hand, the latewood is very dense and is deep-colored, presenting a very decided contrast to the soft, straw-colored earlywood.

It is not only the proportion of latewood, but also its quality, that counts. In specimens that show a very large proportion of latewood it may be noticeably more porous and weigh considerably less than the latewood in pieces that contain but little. One can judge comparative density, and therefore to some extent strength, by visual inspection.

No satisfactory explanation can as yet be given for the exact mechanisms determining the formation of earlywood and latewood. Several factors may be involved. In conifers, at least, rate of growth alone does not determine the proportion of the two portions of the ring, for in some cases the wood of slow growth is very hard and heavy, while in others the opposite is true. The quality of the site where the tree grows undoubtedly affects the character of the wood formed, though it is not possible to formulate a rule governing it. In general, however, it may be said that where strength or ease of working is essential, woods of moderate to slow growth should be chosen.

In Ring-porous Woods

In ring-porous woods each season's growth is always well defined, because the large pores formed early in the season abut on the denser tissue of the year before.

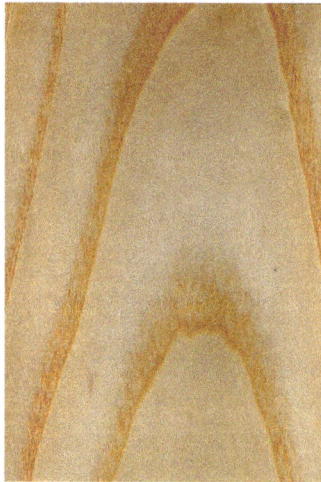

Earlywood and latewood in a ring-porous wood (ash) in a Fraxinus excelsior; tangential view, wide growth rings

In the case of the ring-porous hardwoods there seems to exist a pretty definite relation between the rate of growth of timber and its properties. This may be briefly summed up in the general statement that the more rapid the growth or the wider the rings of growth, the heavier, harder, stronger, and stiffer the wood. This, it must be remembered, applies only to ring-porous woods such as oak, ash, hickory, and others of the same group, and is, of course, subject to some exceptions and limitations.

In ring-porous woods of good growth it is usually the latewood in which the thick-walled, strength-giving fibers are most abundant. As the breadth of ring diminishes, this latewood is reduced so that very slow growth produces comparatively light, porous wood composed of thin-walled vessels and wood parenchyma. In good oak these large vessels of the earlywood occupy from 6 to 10 percent of the volume of the log, while in inferior material they may make up 25% or more. The latewood of good oak is dark colored and firm, and consists mostly of thick-walled fibers which form one-half or more of the wood. In inferior oak, this latewood is much reduced both in quantity and quality. Such variation is very largely the result of rate of growth.

Wide-ringed wood is often called "second-growth", because the growth of the young timber in open stands after the old trees have been removed is more rapid than in trees in a closed forest, and in the manufacture of articles where strength is an important consideration such "second-growth" hardwood material is preferred. This is particularly the case in the choice of hickory for handles and spokes. Here not only strength, but toughness and resilience are important. The results of a series of tests on hickory by the U.S. Forest Service show that:

> "The work or shock-resisting ability is greatest in wide-ringed wood that has from 5 to 14 rings per inch (rings 1.8-5 mm thick), is fairly constant from 14 to 38 rings per inch (rings 0.7–1.8 mm thick), and decreases rapidly from 38 to 47 rings per inch (rings 0.5–0.7 mm thick). The strength at maximum load is not so great with the most rapid-growing wood; it is maximum with from 14 to 20 rings per inch (rings 1.3–1.8 mm thick), and again becomes less as the wood becomes closely ringed. The natural deduction is that wood of first-class mechanical value shows from 5 to 20 rings per inch (rings 1.3–5 mm thick) and that slower growth yields poorer stock. Thus the inspector or buyer of hickory should discriminate against timber that has more than 20 rings per inch (rings less than 1.3 mm thick). Exceptions exist, however, in the case of normal growth upon dry situations, in which the slow-growing material may be strong and tough."

The effect of rate of growth on the qualities of chestnut wood is summarized by the same authority as follows:

> "When the rings are wide, the transition from spring wood to summer wood is gradual, while in the narrow rings the spring wood passes into summer wood abruptly. The width of the spring wood changes but little with the width of the annual ring, so that the narrowing or broadening of the annual ring is always at the expense of the summer wood. The narrow vessels of the summer wood make it richer in wood substance than the spring wood composed of wide vessels. Therefore, rapid-growing specimens with wide rings have more wood substance than slow-growing trees with narrow rings. Since the more the wood substance the greater the weight, and the greater the weight the stronger the wood, chestnuts with wide rings must have stronger wood than chestnuts with narrow rings. This agrees with the accepted view that sprouts (which always have wide rings) yield better and stronger wood than seedling chestnuts, which grow more slowly in diameter."

In Diffuse-porous Woods

In the diffuse-porous woods, the demarcation between rings is not always so clear and in some cases is almost (if not entirely) invisible to the unaided eye. Conversely, when there is a clear demarcation there may not be a noticeable difference in structure within the growth ring.

In diffuse-porous woods, as has been stated, the vessels or pores are even-sized, so that the water conducting capability is scattered throughout the ring instead of collected in the earlywood. The effect of rate of growth is, therefore, not the same as in the ring-porous woods, approaching more nearly the conditions in the conifers. In general it may be stated that such woods of medium growth afford stronger material than when very rapidly or very slowly grown. In many uses of wood, total strength is not the main consideration. If ease of working is prized, wood should be chosen with regard to its uniformity of texture and straightness of grain, which will in most cases occur when there is little contrast between the latewood of one season's growth and the earlywood of the next.

Monocot Wood

Structural material that resembles ordinary, "dicot" or conifer wood in its gross handling characteristics is produced by a number of monocot plants, and these also are colloquially called wood. Of these, bamboo, botanically a member of the grass family, has considerable economic importance, larger culms being widely used as a building and construction material in their own right and, these days, in the manufacture of engineered flooring, panels and veneer. Another major plant group that produce material that often is called wood are the palms. Of much less importance are plants such as *Pandanus, Dracaena* and *Cordyline*. With all this material, the structure and composition of the structural material is quite different from ordinary wood.

Trunks of the coconut palm, a monocot, in Java. From this perspective these look not much different from trunks of a dicot or conifer

Specific Gravity

The single most revealing property of wood as an indicator of wood quality is specific gravity (Timell 1986), as both pulp yield and lumber strength are determined by it. Specific gravity is the ratio of the mass of a substance to the mass of an equal volume of water; density is the ratio of a mass of a quantity of a substance to the volume of that quantity and is expressed in mass per unit substance, e.g., grams per milliliter (g/cm^3 or g/ml). The terms are essentially equivalent as long as the metric system is used. Upon drying, wood shrinks and its density increases. Minimum values are associated with green (water-saturated) wood and are referred to as *basic specific gravity* (Timell 1986).

Wood Density

Wood density is determined by multiple growth and physiological factors compounded into "one fairly easily measured wood characteristic" (Elliott 1970).

Age, diameter, height, radial growth, geographical location, site and growing conditions, silvicultural treatment, and seed source, all to some degree influence wood density. Variation is to be expected. Within an individual tree, the variation in wood density is often as great as or even greater than that between different trees (Timell 1986). Variation of specific gravity within the bole of a tree can occur in either the horizontal or vertical direction.

Hard and Soft Woods

It is common to classify wood as either softwood or hardwood. The wood from conifers (e.g. pine) is called softwood, and the wood from dicotyledons (usually broad-leaved trees, e.g. oak) is called hardwood. These names are a bit misleading, as hardwoods are not necessarily hard, and softwoods are not necessarily soft. The well-known balsa (a hardwood) is actually softer than any commercial softwood. Conversely, some softwoods (e.g. yew) are harder than many hardwoods.

There is a strong relationship between the properties of wood and the properties of the particular tree that yielded it. The density of wood varies with species. The density of a wood correlates with its strength (mechanical properties). For example, mahogany is a medium-dense hardwood that is excellent for fine furniture crafting, whereas balsa is light, making it useful for model building. One of the densest woods is black ironwood.

Chemistry of Wood

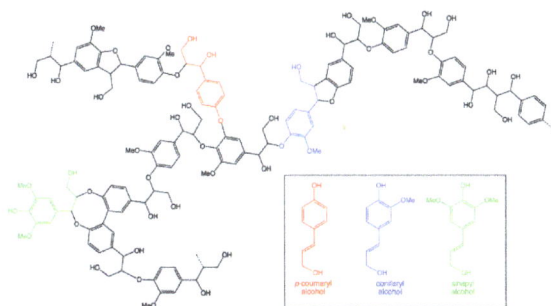

Chemical structure of lignin, which comprises approximately 30% of wood and is responsible for many of its properties.

The chemical composition of wood varies from species to species, but is approximately 50% carbon, 42% oxygen, 6% hydrogen, 1% nitrogen, and 1% other elements (mainly calcium, potassium, sodium, magnesium, iron, and manganese) by weight. Wood also contains sulfur, chlorine, silicon, phosphorus, and other elements in small quantity.

Aside from water, wood has three main components. Cellulose, a crystalline polymer derived from glucose, constitutes about 41–43%. Next in abundance is hemicellulose, which is around 20% in deciduous trees but near 30% in conifers. It is mainly five-carbon sugars that are linked in an irregular manner, in contrast to the cellulose. Lignin is the third component at around 27% in coniferous wood vs. 23% in deciduous trees. Lignin confers the hydrophobic properties reflecting the fact that it is based on aromatic rings. These three components are interwoven, and direct covalent linkages exist between the lignin and the hemicellulose. A major focus of the paper industry is the separation of the lignin from the cellulose, from which paper is made.

In chemical terms, the difference between hardwood and softwood is reflected in the composition of the constituent lignin. Hardwood lignin is primarily derived from sinapyl alcohol and coniferyl alcohol. Softwood lignin is mainly derived from coniferyl alcohol.

Extractives

Aside from the lignocellulose, wood consists of a variety of low molecular weight organic compounds, called *extractives*. The wood extractives are fatty acids, resin acids, waxes and terpenes. For example, rosin is exuded by conifers as protection from insects. The extraction of these organic materials from wood provides tall oil, turpentine, and rosin.

Uses

Fuel

Wood has a long history of being used as fuel, which continues to this day, mostly in rural areas of the world. Hardwood is preferred over softwood because it creates less smoke and burns longer. Adding a woodstove or fireplace to a home is often felt to add ambiance and warmth.

Construction

Wood has been an important construction material since humans began building shelters, houses and boats. Nearly all boats were made out of wood until the late 19th century, and wood remains in common use today in boat construction. Elm in particular was used for this purpose as it resisted decay as long as it was kept wet (it also served for water pipe before the advent of more modern plumbing).

Wood to be used for construction work is commonly known as *lumber* in North America. Elsewhere, *lumber* usually refers to felled trees, and the word for sawn planks ready for use is *timber*. In Medieval Europe oak was the wood of choice for all wood construction, including beams, walls, doors, and floors. Today a wider variety of woods is used: solid wood doors are often made from poplar, small-knotted pine, and Douglas fir.

The Saitta House, Dyker Heights, Brooklyn, New York built in 1899 is made of and decorated in wood.

The churches of Kizhi, Russia are among a handful of World Heritage Sites built entirely of wood, without metal joints.

New domestic housing in many parts of the world today is commonly made from timber-framed construction. Engineered wood products are becoming a bigger part of the construction industry. They may be used in both residential and commercial buildings as structural and aesthetic materials.

In buildings made of other materials, wood will still be found as a supporting material, especially in roof construction, in interior doors and their frames, and as exterior cladding.

Wood is also commonly used as shuttering material to form the mould into which concrete is poured during reinforced concrete construction.

Wood Flooring

Wood can be cut into straight planks and made into a wood flooring.

Engineered Wood

Engineered wood products, glued building products "engineered" for application-specific performance requirements, are often used in construction and industrial applications. Glued engineered wood products are manufactured by bonding together wood strands, veneers, lumber or other forms of wood fiber with glue to form a larger, more efficient composite structural unit.

These products include glued laminated timber (glulam), wood structural panels (including plywood, oriented strand board and composite panels), laminated veneer lumber (LVL) and other structural composite lumber (SCL) products, parallel strand lumber, and I-joists. Approximately 100 million cubic meters of wood was consumed for this purpose in 1991. The trends suggest that particle board and fiber board will overtake plywood.

Wood unsuitable for construction in its native form may be broken down mechanically (into fibers or chips) or chemically (into cellulose) and used as a raw material for other building materials, such as engineered wood, as well as chipboard, hardboard, and medium-density fiberboard (MDF). Such wood derivatives are widely used: wood fibers are an important component of most paper, and cellulose is used as a component of some synthetic materials. Wood derivatives can also be used for kinds of flooring, for example laminate flooring.

Furniture and Utensils

Wood has always been used extensively for furniture, such as chairs and beds. It is also used for tool handles and cutlery, such as chopsticks, toothpicks, and other utensils, like the wooden spoon.

Next Generation Wood Products

Further developments include new lignin glue applications, recyclable food packaging, rubber tire replacement applications, anti-bacterial medical agents, and high strength fabrics or composites. As scientists and engineers further learn and develop new techniques to extract various components from wood, or alternatively to modify wood, for example by adding components to wood, new more advanced products will appear on the marketplace. Moisture content electronic monitoring can also enhance next generation wood protection.

In The Arts

Wood has long been used as an artistic medium. It has been used to make sculptures and carvings for millennia. Examples include the totem poles carved by North American indigenous people from conifer trunks, often Western Red Cedar (*Thuja plicata*), and the Millennium clock tower, now housed in the National Museum of Scotland in Edinburgh. It is also used in woodcut printmaking, and for engraving.

Certain types of musical instruments, such as those of the violin family, the guitar, the clarinet and recorder, the xylophone, and the marimba, are traditionally made mostly or entirely of wood. The choice of wood may make a significant difference to the tone and resonant qualities of the instrument, and tonewoods have widely differing properties, ranging from the hard and dense african blackwood (used for the bodies of clarinets) to the light but resonant European spruce (*Picea abies*), which is traditionally used for the soundboards of violins. The most valuable tonewoods, such as the ripple sycamore (*Acer pseudoplatanus*), used for the backs of violins, combine acoustic properties with decorative color and grain which enhance the appearance of the finished instrument.

Stringed instrument bows are often made from brazilwood (also called pernambuco).

Despite their collective name, not all woodwind instruments are made entirely of wood. The reeds used to play them, however, are usually made from *Arundo donax*, a type of monocot cane plant.

Sports and Recreational Equipment

Many types of sports equipment are made of wood, or were constructed of wood in the past. For example, cricket bats are typically made of white willow. The baseball bats which are legal for use in Major League Baseball are frequently made of ash wood or hickory, and in recent years have been constructed from maple even though that wood is somewhat more fragile. NBA courts have been traditionally made out of parquetry.

Many other types of sports and recreation equipment, such as skis, ice hockey sticks, lacrosse sticks and archery bows, were commonly made of wood in the past, but have since been replaced with more modern materials such as aluminium, fiberglass, carbon fiber, titanium, and composite materials. One noteworthy example of this trend is the golf club commonly known as the *wood*, the head of which was traditionally made of persimmon wood in the early days of the game of golf, but is now generally made of synthetic materials.

Bacterial Degradation

Little is known about the bacteria that degrade cellulose. Symbiotic bacteria in Xylophaga may play a role in the degradation of sunken wood; while bacteria such as Alphaproteobacteria, Flavobacteria, Actinobacteria, Clostridia, and Bacteroidetes have been detected in wood submerged over a year.

Straw

Straw is an agricultural by-product, the dry stalks of cereal plants, after the grain and chaff have been removed. Straw makes up about half of the yield of cereal crops such as barley, oats, rice, rye and wheat. It has many uses, including fuel, livestock bedding and fodder, thatching and basket-making. It is usually gathered and stored in a straw bale, which is a bundle of straw tightly bound with twine or wire. Bales may be square, rectangular, or round, depending on the type of baler used.

Bundles of rice straw

Large round bales

Straw or hay briquettes are a biofuel substitute to coal

Pile of "small square" straw bales, sheltered under a tarpaulin

Uses

Current and historic uses of straw include:

- Animal feed

    o Straw may be fed as part of the roughage component of the diet to cattle or horses that are on a near maintenance level of energy requirement. It has a low digestible energy and nutrient content. The heat generated when microorganisms in a herbivore's gut digest straw can be useful in maintaining body temperature in cold climates. Due to the risk of impaction and its poor nutrient profile, it should always be restricted to part of the diet. It may be fed as it is, or chopped into short lengths, known as chaff.

- Basketry

    o Bee skeps and linen baskets are made from coiled and bound together continuous lengths of straw. The technique is known as lip work.

- Bedding: humans or livestock

    o The straw-filled mattress, also known as a palliasse, is still used in many parts of the world.

    o It is commonly used as bedding for ruminants and horses. It may be used as bedding and food for small animals, but this often leads to injuries to mouth, nose and eyes as straw is quite sharp.

- Biofuels

    o The use of straw as a carbon-neutral energy source is increasing rapidly, especially for biobutanol. Straw or hay briquettes are a biofuel substitute to coal.

- Biogas

    o Straw has been tested for use in a biogas plant in Aarhus University, Denmark. Straw has been processed as briquettes and feed into a biogas plant to see if higher gas yields were attainable.

- Biomass

    o The use of straw in large-scale biomass power plants is becoming mainstream in the EU, with several facilities already online. The straw is either used directly in the form of bales, or densified into pellets which allows for the feedstock to be transported over longer distances. Finally, torrefaction of straw with pelletisation is gaining attention, because it increases the energy density of the resource, making it possible to transport it still further. This processing step also makes storage much easier, because torrefied straw pellets are hydrophobic. Torrefied straw in the form of pellets can be directly co-fired with coal or natural gas at very high rates and make use of the processing infrastructures at existing coal and gas plants. Because the torrefied straw pellets have superior structural, chemical and combustion properties to coal, they can replace all coal and turn a coal plant into an entirely biomass-fed power station. First generation pellets are limited to a co-firing rate of 15% in modern IGCC plants.

- Construction material:

  o In many parts of the world, straw is used to bind clay and concrete. A mixture of clay and straw, known as cob, can be used as a building material. There are many recipes for making cob.

  o When baled, straw has moderate insulation characteristics (about R-1.5/inch according to Oak Ridge National Lab and Forest Product Lab testing). It can be used, alone or in a post-and-beam construction, to build straw bale houses. When bales are used to build or insulate buildings, the straw bales are commonly finished with earthen plaster. The plastered walls provide some thermal mass, compressive and ductile structural strength, and acceptable fire resistance as well as thermal resistance (insulation), somewhat in excess of North American building code. Straw is an abundant agricultural waste product, and requires little energy to bale and transport for construction. For these reasons, strawbale construction is gaining popularity as part of passive solar and other renewable energy projects.

  o Composite lumber Wheat straw can be used as a polymer filler combined with polymers to produce composite lumber.

  o Enviroboard can be made from straw.

  o Strawblocks

- Crafts

Belarusian Straw Dolls

Easter bunny made of Straw

- o   Corn dollies

- o   Straw marquetry

- o   Straw painting

- o   Straw plaiting

- o   Scarecrows

- o   Japanese Traditional Cat's House

- Erosion control

  - o   Straw bales are sometimes used for sediment control at construction sites. However, bales are often ineffective in protecting water quality and are maintenance-intensive. For these reasons the U.S. Environmental Protection Agency (EPA) and various state agencies recommend use of alternative sediment control practices where possible, such as silt fences, fiber rolls and geotextiles.

  - o   Burned area emergency response

  - o   Ground cover

  - o   In-stream check dams

- Hats

  - o   There are several styles of straw hats that are made of woven straw.

  - o   Many thousands of women and children in England (primarily in the Luton district of Bedfordshire), and large numbers in the United States (mostly Massachusetts), were employed in plaiting straw for making hats. By the late 19th century, vast quantities of plaits were being imported to England from Canton in China, and in the United States most of the straw plait was imported.

- o A fiber analogous to straw is obtained from the plant *Carludovica palmata*, and is used to make Panama hats.

- Horticulture

  - o Straw is used in cucumber houses and for mushroom growing.

  - o In Japan, certain trees are wrapped with straw to protect them from the effects of a hard winter as well as to use them as a trap for parasite insects.

  - o It is also used in ponds to reduce algae by changing the nutrient ratios in the water.

  - o The soil under strawberries is covered with straw to protect the ripe berries from dirt, and straw is also used to cover the plants during winter to prevent the cold from killing them.

  - o Straw also makes an excellent mulch.

- Packaging

  - o Straw is resistant to being crushed and therefore makes a good packing material. A company in France makes a straw mat sealed in thin plastic sheets.

  - o Straw envelopes for wine bottles have become rarer, but are still to be found at some wine merchants.

  - o Wheat straw is also used in compostable food packaging such as compostable plates. Packaging made from wheat straw can be certified compostable and will biodegrade in a commercial composting environment.

- Paper

  - o Straw can be pulped to make paper.

Belarusian Straw Bird

- Rope

  o Rope made from straw was used by thatchers, in the packaging industry and even in iron foundries.

- Shoes

  o Koreans wear Jipsin, sandals made of straw.

  o In some parts of Germany like Black Forest and Hunsrück people wear straw shoes at home or at carnival.

- Targets

  o Heavy gauge straw rope is coiled and sewn tightly together to make archery targets. This is no longer done entirely by hand, but is partially mechanised. Sometimes a paper or plastic target is set up in front of straw bales, which serve to support the target and provide a safe backdrop.

- Thatching

  o Thatching uses straw, reed or similar materials to make a waterproof, light-weight roof with good insulation properties. Straw for this purpose (often wheat straw) is grown specially and harvested using a reaper-binder.

Safety

Dried straw presents a fire hazard that can ignite easily if exposed to sparks or an open flame. It can also trigger Allergic rhinitis in people who are hypersensitive to airborne allergens such as straw dust.

Research

In addition to its current and historic uses, straw is being investigated as a source of fine chemicals including alkaloids, flavonoids, lignins, phenols, and steroids.

Manure

Manure is organic matter, mostly derived from animal feces except in the case of green manure, which can be used as organic fertilizer in agriculture. Manures contribute to the fertility of the soil by adding organic matter and nutrients, such as nitrogen, that are trapped by bacteria in the soil. Higher organisms then feed on the fungi and bacteria in a chain of life that comprises the soil food web. It is also a product obtained after decomposition of organic matter like cow dung which replenishes the soil with essential elements and add humus to the soil.

Animal manure is often a mixture of animal feces and bedding straw, as in this example from a stable

In the past, the term "manure" included inorganic fertilizers, but this usage is now very rare.

Types

There are three main classes of manures used in soil management:

Animal Manure

Most animal manure consist of feces. Common forms of animal manure include farmyard manure (FYM) or farm slurry (liquid manure). FYM also contains plant material (often straw), which has been used as bedding for animals and has absorbed the feces and urine. Agricultural manure in liquid form, known as slurry, is produced by more intensive livestock rearing systems where concrete or slats are used, instead of straw bedding. Manure from different animals has different qualities and requires different application rates when used as fertilizer. For example horses, cattle, pigs, sheep, chickens, turkeys, rabbits, and guano from seabirds and bats all have different properties. For instance, sheep manure is high in nitrogen and potash, while pig manure is relatively low in both. Horses mainly eat grass and a few weeds so horse manure can contain grass and weed seeds, as horses do not digest seeds the way that cattle do. Chicken litter, coming from a bird, is very concentrated in nitrogen and phosphate and is prized for both properties.

Cement reservoirs, one new, and one containing cow manure mixed with water. This is common in rural Hainan Province, China.

Animal manures may be adulterated or contaminated with other animal products, such as wool (shoddy and other hair), feathers, blood, and bone. Livestock feed can be mixed with the manure due to spillage. For example, chickens are often fed meat and bone meal, an animal product, which can end up becoming mixed with chicken litter.

Human Manure

Some people refer to human excreta as human manure, and the word "humanure" has also been used. Just like animal manure, it can be applied as a soil conditioner (reuse of excreta in agriculture). Sewage sludge is a material that contains human excreta, as it is generated after mixing excreta with water and treatment of the wastewater in a sewage treatment plant.

Compost containing turkey manure and wood chips from bedding material is dried and then applied to pastures for fertilizer.

Compost

Compost is the decomposed remnants of organic materials. It is usually of plant origin, but often includes some animal dung or bedding.

Green Manure

Green manures are crops grown for the express purpose of plowing them in, thus increasing fertility through the incorporation of nutrients and organic matter into the soil. Leguminous plants such as clover are often used for this, as they fix nitrogen using *Rhizobia* bacteria in specialized nodes in the root structure.

Other types of plant matter used as manure include the contents of the rumens of slaughtered ruminants, spent grain (left over from brewing beer) and seaweed.

Uses of Manure

Animal Manure

Animal manure, such as chicken manure and cow dung, has been used for centuries as a fertilizer for farming. It can improve the soil structure (aggregation) so that the soil holds more nutrients and water, and therefore becomes more fertile. Animal manure also encourages soil microbial activity which promotes the soil's trace mineral supply, improving plant nutrition. It also contains some nitrogen and other nutrients that assist the growth of plants.

Manure on a wall.

Manures with a particularly unpleasant odor (such as slurries from intensive pig farming) are usually knifed (injected) directly into the soil to reduce release of the odor. Manure from pigs and cattle is usually spread on fields using a manure spreader. Due to the relatively lower level of proteins in vegetable matter, herbivore manure has a milder smell than the dung of carnivores or omnivores. However, herbivore slurry that has undergone anaerobic fermentation may develop more unpleasant odors, and this can be a problem in some agricultural regions. Poultry droppings are harmful to plants when fresh but, after a period of composting, are valuable fertilizers.

Manure is also commercially composted and bagged and sold retail as a soil amendment.

Before motor vehicles became common, horse droppings were a big part of the rubbish that communities needed to clean off roads.

Precautions

Manure generates heat as it decomposes, and it is possible for manure to ignite spontaneously if stored in a very large pile. Once such a large pile of manure is burning, it will foul the air over a wide area and require considerable effort to extinguish. Therefore, large feedlots must take care to ensure that piles of fresh manure do not get excessively large. There is no serious risk of spontaneous combustion in smaller operations.

There is also a risk of insects carrying feces to food and water supplies, making them unsuitable for human consumption.

Livestock Antibiotics

In 2007, a University of Minnesota study indicated that foods such as corn, lettuce, and potatoes have been found to accumulate antibiotics from soils spread with animal manure that contains these drugs.

Organic foods may be much more or much less likely to contain antibiotics, depending on their sources and treatment of manure. For instance, by Soil Association Standard 4.7.38, most organic arable farmers either have their own supply of manure (which would, therefore, not normally contain drug residues) or else rely on green manure crops for the extra fertility (if any nonorganic manure is used by organic farmers, then it usually has to be rotted or composted to degrade any residues of drugs and eliminate any pathogenic bacteria — Standard 4.7.38, Soil Association organic farming standards). On the other hand, as found in the University of Minnesota study, the non-usage of artificial fertilizers, and resulting exclusive use of manure as fertilizer, by organic farmers can result in significantly greater accumulations of antibiotics in organic foods.

Sugarcane

Sugar cane growing, Queensland, 2016   Saccharum officinarum

Sugarcane, or sugar cane, is one of the several species of tall perennial true grasses of the genus *Saccharum*, tribe Andropogoneae, native to the warm temperate to tropical regions of South Asia and Melanesia, and used for sugar production. It has stout jointed fibrous stalks that are rich in the sugar sucrose, which accumulates in the stalk internodes. The plant is two to six metres (six feet seven inches to nineteen feet eight inches) tall. All sugar cane species interbreed and the major commercial cultivars are complex hybrids. Sugarcane belongs to the grass family Poaceae, an economically important seed plant family that includes maize, wheat, rice, and sorghum and many forage crops.

Sugarcane and bowl of refined sugar

Cut sugarcane

Sucrose, extracted and purified in specialized mill factories, is used as raw material in human food industries or is fermented to produce ethanol. Ethanol is produced on a large scale by the Brazilian sugarcane industry. Sugarcane is the world's largest crop by production quantity. In 2012, The Food and Agriculture Organization (FAO) estimates it was cultivated on about 26×10^6 hectares (6.425×10^7 acres), in more than 90 countries, with a worldwide harvest of 1.83×10^9 tonnes (1.80×10^9 long tons; 2.02×10^9 short tons). Brazil was the largest producer of sugar cane in the world. The next five major producers, in decreasing amounts of production, were India, China, Thailand, Pakistan and Mexico.

The world demand for sugar is the primary driver of sugarcane agriculture. Cane accounts for 80% of sugar produced; most of the rest is made from sugar beets. Sugarcane predominantly grows in the tropical and subtropical regions (sugar beets grow in colder temperate regions). Other than sugar, products derived from sugarcane include falernum, molasses, rum, *cachaça* (a traditional spirit from Brazil), bagasse and ethanol. In some regions, people use sugarcane reeds to make pens, mats, screens, and thatch. The young unexpanded inflorescence of *tebu telor* is eaten raw, steamed or toasted, and prepared in various ways in certain island communities of Indonesia.

The Persians, followed by the Greeks, discovered the famous "reeds that produce honey without bees" in India between the 6th and 4th centuries BC. They adopted and then spread sugarcane agriculture. Merchants began to trade in sugar from India, which was considered a luxury and an expensive spice. In the 18th century AD, sugarcane plantations began in Caribbean, South American, Indian Ocean and Pacific island nations and the need for laborers became a major driver of large human migrations, including slave labor and indentured servants.

Description

Sugarcane is a tropical, perennial grass that forms lateral shoots at the base to produce multiple stems, typically three to four metres (9 feet 10 inches to 13 feet 1 inch) high and about 5 cm (2 inches) in diameter. The stems grow into cane stalk, which when mature constitutes approximately 75% of the entire plant. A mature stalk is typically composed of 11–16% fiber, 12–16% soluble sugars, 2–3% non-sugars, and 63–73% water. A sugarcane crop is sensitive to the climate, soil type, irrigation, fertilizers, insects, disease control, varieties, and the harvest period. The average yield of cane stalk is 60–70 tonnes per hectare (24–28 long ton/acre; 27–31 short ton/acre) per year. However, this figure can vary between 30 and 180 tonnes per hectare depending on knowledge and crop management approach used in sugarcane cultivation. Sugarcane is a cash crop, but it is also used as livestock fodder.

History

Sugarcane is indigenous to tropical South and Southeast Asia. Different species likely originated in different locations, with *Saccharum barberi* originating in India and *S. edule* and *S. officinarum* in New Guinea. It is theorized that sugarcane was first domesticated as a crop in New Guinea around 6000 BC. New Guinean farmers and other early cultivators of sugarcane chewed the plant for its sweet juice. Early farmers in Southeast Asia, and elsewhere, may have also boiled the cane juice down to a viscous mass to facilitate transportation, but the earliest known production of crystalline sugar began in northern India. The exact date of the first cane sugar production is unclear. The earliest evidence of sugar production comes from ancient Sanskrit and Pali texts.

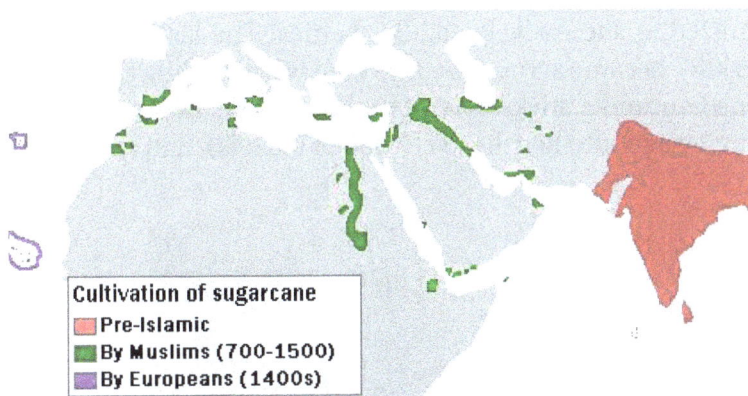

Cultivation of sugarcane
- Pre-Islamic
- By Muslims (700-1500)
- By Europeans (1400s)

The westward diffusion of sugarcane in pre-Islamic times (shown in red), in the medieval Muslim world (green) and by Europeans in the 15th century (islands circled by violet lines)

Around the 8th century, Arab traders introduced sugar from South Asia to the other parts of the Abbasid Caliphate in the Mediterranean, Mesopotamia, Egypt, North Africa, and Andalusia. By the 10th century, sources state that there was no village in Mesopotamia that did not grow sugarcane. It was among the early crops brought to the Americas by the Spanish, mainly Andalusians, from their fields in the Canary Islands, and the Portuguese from their fields in the Madeira Islands.

Christopher Columbus first brought sugarcane to the Caribbean during his second voyage to the Americas; initially to the island of Hispaniola (modern day Haiti and the Dominican Republic). In colonial times, sugar formed one side of the triangle trade of New World raw materials, along with European manufactured goods, and African slaves. Sugar (often in the form of molasses) was shipped from the Caribbean to Europe or New England, where it was used to make rum. The profits from the sale of sugar were then used to purchase manufactured goods, which were then shipped to West Africa, where they were bartered for slaves. The slaves were then brought back to the Caribbean to be sold to sugar planters. The profits from the sale of the slaves were then used to buy more sugar, which was shipped to Europe.

France found its sugarcane islands so valuable that it effectively traded its portion of Canada, famously dubbed "a few acres of snow", to Britain for their return of Guadeloupe, Martinique and St. Lucia at the end of the Seven Years' War. The Dutch similarly kept Suriname, a sugar colony in South America, instead of seeking the return of the New Netherlands (New York).

Boiling houses in the 17th through 19th centuries converted sugarcane juice into raw sugar. These houses were attached to sugar plantations in the Western colonies. Slaves often ran the boiling process under very poor conditions. Rectangular boxes of brick or stone served as furnaces, with an opening at the bottom to stoke the fire and remove ashes. At the top of each furnace were up to seven copper kettles or boilers, each one smaller and hotter than the previous one. The cane juice began in the largest kettle. The juice was then heated and lime added to remove impurities. The juice was skimmed and then channeled to successively smaller kettles. The last kettle, the "teache", was where the cane juice became syrup. The next step was a cooling trough, where the sugar crystals hardened around a sticky core of molasses. This raw sugar was then shoveled from the cooling trough into hogsheads (wooden barrels), and from there into the curing house.

A sugar plantation on the island of Réunion in the late 19th century

In the British Empire, slaves were liberated after 1833 and many would no longer work on sugar cane plantations when they had a choice. British owners of sugar cane plantations therefore needed new workers, and they found cheap labour in China, Portugal and India. The people were subject to indenture, a long-established form of contract which bound them to forced labour for a fixed term; apart from the fixed term of servitude, this resembled slavery. The first ships carrying indentured labourers from India left in 1836. The migrations to serve sugarcane plantations led to a significant number of ethnic Indians, southeast Asians and Chinese settling in various parts of the world. In some islands and countries, the South Asian migrants now constitute between 10 to

50 percent of the population. Sugarcane plantations and Asian ethnic groups continue to thrive in countries such as Fiji, Natal, Burma, Sri Lanka, Malaysia, British Guiana, Jamaica, Trinidad, Martinique, French Guiana, Guadeloupe, Grenada, St. Lucia, St. Vincent, St. Kitts, St. Croix, Suriname, Nevis, Mauritius.

Old-fashioned Indian Sugarcane Press, c. 1905

The then British colony of Queensland, now a state of Australia, imported between 55,000 and 62,500 (estimates vary) people from the South Pacific Islands to work on sugarcane plantations between 1863 and 1900.

Cuban sugar derived from sugarcane was exported to the USSR where it received price supports and was ensured a guaranteed market. The 1991 dissolution of the Soviet state forced the closure of most of Cuba's sugar industry.

Sugarcane remains an important part of the economy of Guyana, Belize, Barbados and Haiti, along with the Dominican Republic, Guadeloupe, Jamaica, and other islands.

Approximately 70% of the sugar produced globally comes from *S. officinarum* and hybrids using this species.

A 19th-century lithograph by Theodore Bray showing a sugarcane plantation. On right is "white officer", the European overseer. Slave workers toil during the harvest. To the left is a flat-bottomed vessel for cane transportation.

Cultivation

Sugarcane cultivation requires a tropical or temperate climate, with a minimum of 60 centimetres (24 in) of annual moisture. It is one of the most efficient photosynthesizers in the plant kingdom. It is a C_4 plant, able to convert up to one percent of incident solar energy into biomass. In prime growing regions, such as Mauritius, Dominican Republic, Puerto Rico, India, Guyana, Indonesia, Pakistan, Peru, Brazil, Bolivia, Colombia, Australia, Ecuador, Cuba, the Philippines, El Salvador, Jamaica and Hawaii, sugarcane crops can produce over 15 kilograms of cane per square metre. Once a major crop of the southeastern region of the United States, sugarcane cultivation has declined there in recent decades, and is now primarily confined to Florida and Louisiana.

Sugarcane is cultivated in the tropics and subtropics in areas with a plentiful supply of water, for a continuous period of more than six to seven months each year, either from natural rainfall or through irrigation. The crop does not tolerate severe frosts. Therefore, most of the world's sugarcane is grown between 22°N and 22°S, and some up to 33°N and 33°S. When sugarcane crop is found outside this range, such as the Natal region of South Africa, it is normally due to anomalous climatic conditions in the region, such as warm ocean currents that sweep down the coast. In terms of altitude, sugarcane crop is found up to 1,600 m close to the equator in countries such as Colombia, Ecuador and Peru.

Sugarcane plantation in Mauritius

Sugarcane can be grown on many soils ranging from highly fertile well drained mollisols, through heavy cracking vertisols, infertile acid oxisols, peaty histosols to rocky andisols. Both plentiful sunshine and water supplies increase cane production. This has made desert countries with good irrigation facilities such as Egypt as some of the highest yielding sugarcane cultivating regions.

Sugarcane flower in Dominica.

Although sugarcanes produce seeds, modern stem cutting has become the most common reproduction method. Each cutting must contain at least one bud, and the cuttings are sometimes hand-planted. In more technologically advanced countries like the United States and Australia, billet planting is common. Billets harvested from a mechanical harvester are planted by a machine that opens and recloses the ground. Once planted, a stand can be harvested several times; after each harvest, the cane sends up new stalks, called ratoons. Successive harvests give decreasing yields, eventually justifying replanting. Two to 10 harvests are usually made depending on the type of culture. In a country with a mechanical agriculture looking for a high production of large fields, like in North America, sugar canes are replanted after two or three harvests to avoid a lowering in yields. In countries with a more traditional type of agriculture with smaller fields and hand harvesting, like in the French island la Réunion, sugar canes are often harvested up to 10 years before replanting.

Sugar canes harvested by women in Hòa Bình province, Vietnam.

Sugarcane is harvested by hand and mechanically. Hand harvesting accounts for more than half of production, and is dominant in the developing world. In hand harvesting, the field is first set on fire. The fire burns dry leaves, and chases away or kills any lurking venomous snakes, without harming the stalks and roots. Harvesters then cut the cane just above ground-level using cane knives or machetes. A skilled harvester can cut 500 kilograms (1,100 lb) of sugarcane per hour.

Sugarcane mechanical harvest in Jaboticabal, São Paulo, Brazil

Mechanical harvesting uses a combine, or sugarcane harvester. The Austoft 7000 series, the original modern harvester design, has now been copied by other companies, including Cameco / John Deere. The machine cuts the cane at the base of the stalk, strips the leaves, chops the cane into consistent lengths and deposits it into a transporter following alongside. The harvester then blows the trash back onto the field. Such machines can harvest 100 long tons (100 t) each hour; however, harvested cane must be rapidly processed. Once cut, sugarcane begins to lose its sugar content, and damage to the cane during mechanical harvesting accelerates this decline. This decline is offset because a modern chopper harvester can complete the harvest faster and more efficiently than hand cutting and loading. Austoft also developed a series of hydraulic high-lift infield transporters to work alongside their harvesters to allow even more rapid transfer of cane to, for example, the nearest railway siding. This mechanical harvesting doesn't require the field to be set on fire; the remains left in the field by the machine consist of the top of the sugar cane and the dead leaves, which act as mulch for the next round of planting.

A panoramic view of sugarcane plantations in Brazil, the largest producer in the world.

Pests

The cane beetle (also known as cane grub) can substantially reduce crop yield by eating roots; it can be controlled with imidacloprid (Confidor) or chlorpyrifos (Lorsban). Other important pests are the larvae of some butterfly/moth species, including the turnip moth, the sugarcane borer (*Diatraea saccharalis*), the Mexican rice borer (*Eoreuma loftini*); leaf-cutting ants, termites, spittlebugs (especially *Mahanarva fimbriolata* and *Deois flavopicta*), and the beetle *Migdolus fryanus*. The planthopper insect *Eumetopina flavipes* acts as a virus vector, which causes the sugarcane disease ramu stunt.

Pathogens

Numerous pathogens infect sugarcane, such as sugarcane grassy shoot disease caused by *Phytoplasma*, whiptail disease or sugarcane smut, *pokkah boeng* caused by *Fusarium moniliforme*, Xanthomonas axonopodis bacteria causes Gumming Disease, and red rot disease caused by *Colletotrichum falcatum*. Viral diseases affecting sugarcane include sugarcane mosaic virus, maize streak virus, and sugarcane yellow leaf virus.

Nitrogen Fixation

Some sugarcane varieties are capable of fixing atmospheric nitrogen in association with the bacterium *Glucoacetobacter diazotrophicus*. Unlike legumes and other nitrogen-fixing plants that form root nodules in the soil in association with bacteria, *G.*

diazotrophicus lives within the intercellular spaces of the sugarcane's stem. Coating seeds with the bacteria is a newly developed technology that can enable every crop species to fix nitrogen for its own use.

Conditions for Sugarcane Workers

At least 20,000 people are estimated to have died of chronic kidney disease (CKD) in Central America in the past two decades – most of them sugar cane workers along the Pacific coast. This may be due to working long hours in the heat without adequate fluid intake.

Processing

Traditionally, sugarcane processing requires two stages. Mills extract raw sugar from freshly harvested cane and "mill-white" sugar is sometimes produced immediately after the first stage at sugar-extraction mills, intended for local consumption. Sugar crystals appear naturally in white color during the crystallization process. Sulfur dioxide is added to inhibit the formation of color-inducing molecules as well as to stabilize the sugar juices during evaporation. Refineries, often located nearer to consumers in North America, Europe, and Japan, then produce refined white sugar, which is 99 percent sucrose. These two stages are slowly merging. Increasing affluence in the sugar-producing tropics increased demand for refined sugar products, driving a trend toward combined milling and refining.

Milling

Sugarcane processing produces cane sugar (sucrose) from sugarcane. Other products of the processing include bagasse, molasses, and filtercake.

Brown (top) and white sugar crystals.

A truck hauls cane to a sugar mill in Florida

Manually extracting juice from sugarcane

Bagasse, the residual dry fiber of the cane after cane juice has been extracted, is used for several purposes:

- fuel for the boilers and kilns,

- production of paper, paperboard products, and reconstituted panelboard,

- agricultural mulch, and

- as a raw material for production of chemicals.

Santa Elisa sugarcane processing plant in Sertãozinho, one of the largest and oldest in Brazil

The primary use of bagasse and bagasse residue is as a fuel source for the boilers in the generation of process steam in sugar plants. Dried filtercake is used as an animal feed supplement, fertilizer, and source of sugarcane wax.

Molasses is produced in two forms: Blackstrap, which has a characteristic strong flavor, and a purer molasses syrup. Blackstrap molasses is sold as a food and dietary supplement. It is also a common ingredient in animal feed, is used to produce ethanol and rum, and in the manufacturing of citric acid. Purer molasses syrups are sold as molasses, and may also be blended with maple syrup, invert sugars, or corn syrup. Both forms of molasses are used in baking.

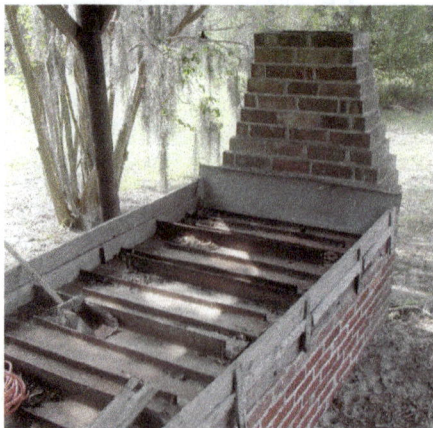

Evaporator with baffled pan and foam dipper for making ribbon cane syrup

Refining

Sugar refining further purifies the raw sugar. It is first mixed with heavy syrup and then centrifuged in a process called "affination". Its purpose is to wash away the sugar crystals' outer coating, which is less pure than the crystal interior. The remaining sugar is then dissolved to make a syrup, about 60 percent solids by weight.

A video of Sugarcane juice extraction

The sugar solution is clarified by the addition of phosphoric acid and calcium hydroxide, which combine to precipitate calcium phosphate. The calcium phosphate particles entrap some impurities and absorb others, and then float to the top of the tank, where they can be skimmed off. An alternative to this "phosphatation" technique is "carbonatation", which is similar, but uses carbon dioxide and calcium hydroxide to produce a calcium carbonate precipitate.

After filtering any remaining solids, the clarified syrup is decolorized by filtration through activated carbon. Bone char or coal-based activated carbon is traditionally used in this role. Some remaining color-forming impurities adsorb to the carbon. The purified syrup is then concentrated to supersaturation and repeatedly crystallized in a vacuum, to produce white refined sugar. As in a sugar mill, the sugar crystals are separated from the molasses by centrifuging. Additional sugar is recovered by blending the remaining syrup with the washings from affination and again crystallizing to produce brown sugar. When no more sugar can be economically recovered, the final molasses still contains 20–30 percent sucrose and 15–25 percent glucose and fructose.

To produce granulated sugar, in which individual grains do not clump, sugar must be dried, first by heating in a rotary dryer, and then by blowing cool air through it for several days.

Ribbon Cane Syrup

Ribbon cane is a subtropical type that was once widely grown in the southern United States, as far north as coastal North Carolina. The juice was extracted with horse or mule-powered crushers; the juice was boiled, like maple syrup, in a flat pan, and then used in the syrup form as a food sweetener. It is not currently a commercial crop, but a few growers find ready sales for their product.

Pollution from Sugarcane Processing

Particulate matter, combustion products, and volatile organic compounds are the primary pollutants emitted during the sugarcane processing. Combustion products include nitrogen oxides (NO_x), carbon monoxide (CO), CO_2, and sulfur oxides (SO_x). Potential emission sources include the sugar granulators, sugar conveying and packaging equipment, bulk loadout operations, boilers, granular carbon and char regeneration kilns, regenerated adsorbent transport systems, kilns and handling equipment (at some facilities), carbonation tanks, multi-effect evaporator stations, and vacuum boiling pans. Modern pollution prevention technologies are capable of addressing all of these potential pollutants.

Brazil led the world in sugarcane production in 2013 with a 739 267 TMT harvest. India was the second largest producer with 341 200 TMT tons, and China the third largest producer with 125 536 TMT tons harvest.

The average worldwide yield of sugarcane crops in 2013 was 70.77 tons per hectare. The most productive farms in the world were in Peru with a nationwide average sugarcane crop yield of 133.71 tons per hectare.

The theoretical possible yield for sugar cane, according to 1983 study of Duke, is about 280 metric tons per hectare per year, and small experimental plots in Brazil have demonstrated yields of 236–280 metric tons of fresh cane per hectare. The most promising region for high yield sugarcane production were in sun drenched, irrigated farms of northern Africa, and other deserts with plentiful water from river or irrigation canals.

In the United States, sugarcane is grown commercially in Florida, Hawaii, Louisiana, and Texas.

Brazil uses sugarcane to produce sugar and ethanol for gasoline-ethanol blends (gasohol), a locally popular transportation fuel. In India, sugarcane is used to produce sugar, jaggery and alcoholic beverages.

Cane Ethanol

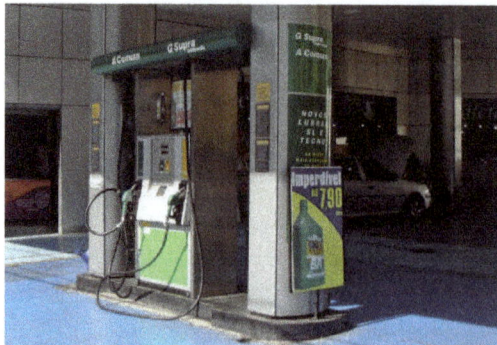

A fuel pump in Brazil, offering cane ethanol (A) and gasoline (G).

Ethanol is generally available as a byproduct of sugar production. It can be used as a biofuel alternative to gasoline, and is widely used in cars in Brazil. It is an alternative to gasoline, and may become the primary product of sugarcane processing, rather than sugar.

In Brazil, gasoline is required to contain at least 22 percent bioethanol. This bioethanol is sourced from Brazil's large sugarcane crop.

The production of ethanol from sugar cane is more energy efficient than from corn or sugar beets or palm/vegetable oils, particularly if cane bagasse is used to produce heat and power for the process. Furthermore, if biofuels are used for crop production and transport, the fossil energy input needed for each ethanol energy unit can be very low. EIA estimates that with an integrated sugar cane to ethanol technology, the well-to-wheels CO_2 emissions can be 90 percent lower than conventional gasoline.

A textbook on renewable energy describes the energy transformation:

Presently, 75 tons of raw sugar cane are produced annually per hectare in Brazil. The cane delivered to the processing plant is called burned and cropped (b&c), and represents 77% of the mass of the raw cane. The reason for this reduction is that the stalks are separated from the leaves (which are burned and whose ashes are left in the field as fertilizer), and from the roots that remain in the ground to sprout for the next crop. Average cane production is, therefore, 58 tons of b&c per hectare per year.

Each ton of b&c yields 740 kg of juice (135 kg of sucrose and 605 kg of water) and 260 kg of moist bagasse (130 kg of dry bagasse). Since the lower heating value of sucrose is 16.5 MJ/kg, and that of the bagasse is 19.2 MJ/kg, the total heating value of a ton of b&c is 4.7 GJ of which 2.2 GJ come from the sucrose and 2.5 from the bagasse.

Per hectare per year, the biomass produced corresponds to 0.27 TJ. This is equivalent to 0.86 W per square meter. Assuming an average insolation of 225 W per square meter, the photosynthetic efficiency of sugar cane is 0.38%.

The 135 kg of sucrose found in 1 ton of b&c are transformed into 70 litres of ethanol with a combustion energy of 1.7 GJ. The practical sucrose-ethanol conversion efficiency is, therefore, 76% (compare with the theoretical 97%).

One hectare of sugar cane yields 4,000 litres of ethanol per year (without any additional energy input, because the bagasse produced exceeds the amount needed to distill the final product). This, however, does not include the energy used in tilling, transportation, and so on. Thus, the solar energy-to-ethanol conversion efficiency is 0.13%.

Bagasse Applications

Sugarcane is a major crop in many countries. It is one of the plants with the highest bioconversion efficiency. Sugarcane crop is able to efficiently fix solar energy, yielding some 55 tonnes of dry matter per hectare of land annually. After harvest, the crop produces sugar juice and bagasse, the fibrous dry matter. This dry matter is biomass with potential as fuel for energy production. Bagasse can also be used as an alternative source of pulp for paper production.

Sugarcane bagasse is a potentially abundant source of energy for large producers of sugarcane, such as Brazil, India and China. According to one report, with use of latest technologies, bagasse produced annually in Brazil has the potential of meeting 20 percent of Brazil's energy consumption by 2020.

Electricity Production

A number of countries, in particular those devoid of any fossil fuel, have implemented energy conservation and efficiency measures to minimize energy used in cane processing and furthermore export any excess electricity to the grid. Bagasse is usually burned to produce steam, which in turn creates electricity. Current technologies, such as those in use in Mauritius, produce over 100 kWh of electricity per tonne of bagasse. With a total world harvest of over 1 billion tonnes of sugar cane per year, the global energy potential from bagasse is over 100,000 GWh. Using Mauritius as a reference, an annual potential of 10,000 GWh of additional electricity could be produced throughout Africa. Electrical generation from bagasse could become quite important, particularly to the rural populations of sugarcane producing nations.

Recent cogeneration technology plants are being designed to produce from 200 to over 300 kWh of electricity per tonne of bagasse. As sugarcane is a seasonal crop, shortly after harvest the supply of bagasse would peak, requiring power generation plants to strategically manage the storage of bagasse.

Biogas Production

A greener alternative to burning bagasse for the production of electricity is to convert bagasse into biogas. Technologies are being developed to use enzymes to transform bagasse into advanced biofuel and biogas.

Sugarcane as Food

Caipirinha, a cocktail made from sugarcane-derived Cachaça

In most countries where sugarcane is cultivated, there are several foods and popular dishes derived directly from it, such as:

- Raw sugarcane: chewed to extract the juice

- *Sayur nganten*: an Indonesian soup made with the stem of trubuk (*Saccharum edule*), a type of sugarcane.

- Sugarcane juice: a combination of fresh juice, extracted by hand or small mills, with a touch of lemon and ice to make a popular drink, known variously as *air tebu, usacha rass, guarab, guarapa, guarapo, papelón, aseer asab, ganna sharbat, mosto, caldo de cana, nᵯᵯc miá.*

- Syrup: a traditional sweetener in soft drinks, now largely supplanted in the US by high fructose corn syrup, which is less expensive because of corn subsidies and sugar tariffs.

- Molasses: used as a sweetener and a syrup accompanying other foods, such as cheese or cookies

- Jaggery: a solidified molasses, known as *gur* or *gud* or *gul* in India, is traditionally produced by evaporating juice to make a thick sludge, and then cooling and molding it in buckets. Modern production partially freeze dries the juice to reduce caramelization and lighten its color. It is used as sweetener in cooking traditional entrees, sweets and desserts.

- Falernum: a sweet, and lightly alcoholic drink made from sugarcane juice

- *Cachaça*: the most popular distilled alcoholic beverage in Brazil; a liquor made of the distillation of sugarcane juice.

- Rum: is a liquor made from sugarcane products, typically molasses but sometimes also cane juice. It is most commonly produced in the Caribbean and environs.

- Basi: is a fermented alcoholic beverage made from sugarcane juice produced in the Philippines and Guyana.

- *Panela*: solid pieces of sucrose and fructose obtained from the boiling and evaporation of sugarcane juice; a food staple in Colombia and other countries in South and Central America

- *Rapadura*: a sweet flour that is one of the simplest refinings of sugarcane juice, common in Latin American countries such as Brazil, Argentina and Venezuela (where it is known as *papelón*) and the Caribbean.

- Rock candy: crystallized cane juice

- Gâteau de Sirop

References

- Jean-Pierre Barette; Claude Hazard et Jérôme Mayer (1996). Mémotech Bois et Matériaux Associés. Paris: Éditions Casteilla. p. 22. ISBN 27135-1645-5.

- Mimms, Agneta; Michael J. Kuckurek; Jef A. Pyiatte; Elizabeth E. Wright (1993). Kraft Pulping. A Compilation of Notes. TAPPI Press. pp. 6–7. ISBN 0-89852-322-2.

- Fiebach, Klemens; Grimm, Dieter (2000). "Resins, Natural". Ullmann's Encyclopedia of Industrial Chemistry. doi:10.1002/14356007.a23_073. ISBN 978-3-527-30673-2.

- Walton Lai (1993). Indentured labor, Caribbean sugar: Chinese and Indian migrants to the British West Indies, 1838–1918. ISBN 978-0-8018-7746-9.

- Steven Vertovik (Robin Cohen, ed.) (1995). The Cambridge survey of world migration. pp. 57–68. ISBN 978-0-521-44405-7.

- K Laurence (1994). A Question of Labour: Indentured Immigration Into Trinidad & British Guiana, 1875–1917. St Martin's Press. ISBN 978-0-312-12172-3.

- Lakhani, Nina (16 February 2015). "Nicaraguans demand action over illness killing thousands of sugar cane workers". The Guardian. Retrieved 2015-04-09.

- Christina Bienhold; Petra Pop Ristova; Frank Wenzhöfer; Thorsten Dittmar; Antje Boetius (January 2, 2013). "How Deep-Sea Wood Falls Sustain Chemosynthetic Life". PLOS ONE.

- "The Photosynthetic Process". Concepts in Photobiology: Photosynthesis and Photomorphogenesis. University of Illinois. Retrieved 2012-04-02.

- "Sugar-Cane Harvester Cuts Forty-Tons an Hour". Popular Mechanics Monthly. Google Books. July 1930. Retrieved 2012-04-02.

- ProfessionalNetSolutions.com. "The Millennium Clock Tower at Edinburgh Royal Museum". Freespace.virgin.net. Retrieved 2011-12-15.

5

Different Types of Efficient Fuels

Efficient fuels are best understood in confluence with the major types listed in the following section. The different types of efficient fuels discussed in this text are biogas, biofuel and pellet fuel. Biogas is the combination of different gases, produced by the breakdown of organic matter in the absence of oxygen. This chapter has been carefully written to provide an easy understanding of the different types of efficient fuels.

Biogas

Biogas typically refers to a mixture of different gases produced by the breakdown of organic matter in the absence of oxygen. Biogas can be produced from raw materials such as agricultural waste, manure, municipal waste, plant material, sewage, green waste or food waste. Biogas is a renewable energy source and in many cases exerts a very small carbon footprint.

Pipes carrying biogas (foreground), natural gas and condensate

Biogas can be produced by anaerobic digestion with anaerobic organisms, which digest material inside a closed system, or fermentation of biodegradable materials.

Biogas is primarily methane (CH4) and carbon dioxide (CO_2) and may have small amounts of hydrogen sulfide (H2S), moisture and siloxanes. The gases methane, hydrogen, and carbon monoxide (CO) can be combusted or oxidized with oxygen. This energy release allows biogas to be used as a fuel; it can be used for any heating purpose, such as cooking. It can also be used in a gas engine to convert the energy in the gas into electricity and heat.

Biogas can be compressed, the same way natural gas is compressed to CNG, and used to power motor vehicles. In the UK, for example, biogas is estimated to have the potential to replace around 17% of vehicle fuel. It qualifies for renewable energy subsidies in some parts of the world. Biogas can be cleaned and upgraded to natural gas standards, when it becomes bio-methane. Biogas is considered to be a renewable resource because its production-and-use cycle is continuous, and it generates no net carbon dioxide. Organic material grows, is converted and used and then regrows in a continually repeating cycle. From a carbon perspective, as much carbon dioxide is absorbed from the atmosphere in the growth of the primary bio-resource as is released when the material is ultimately converted to energy.

Production

Biogas is produced as landfill gas (LFG), which is produced by the breakdown of Biodegradable waste inside a landfill due to chemical reactions and microbes, or as digested gas, produced inside an anaerobic digester. A *biogas plant* is the name often given to an anaerobic digester that treats farm wastes or energy crops. It can be produced using anaerobic digesters (air-tight tanks with different configurations). These plants can be fed with energy crops such as maize silage or biodegradable wastes including sewage sludge and food waste. During the process, the microorganisms transform biomass waste into biogas (mainly methane and carbon dioxide) and digestate. The biogas is a renewable energy that can be used for heating, electricity, and many other operations that use a reciprocating internal combustion engine, such as GE Jenbacher or Caterpillar gas engines. Other internal combustion engines such as gas turbines are suitable for the conversion of biogas into both electricity and heat. The digestate is the remaining inorganic matter that was not transformed into biogas. It can be used as an agricultural fertiliser.

Biogas production in rural Germany

There are two key processes: mesophilic and thermophilic digestion which is dependent on temperature. In experimental work at University of Alaska Fairbanks, a 1000-litre digester using psychrophiles harvested from "mud from a frozen lake in Alaska" has produced 200–300 liters of methane per day, about 20%–30% of the output from digesters in warmer climates.

Dangers

The dangers of biogas are mostly similar to those of natural gas, but with an additional risk from the toxicity of its hydrogen sulfide fraction. Biogas can be explosive when mixed in the ratio of one part biogas to 8-20 parts air. Special safety precautions have to be taken for entering an empty biogas digester for maintenance work.

It is important that a biogas system never has negative pressure as this could cause an explosion. Negative gas pressure can occur if too much gas is removed or leaked; Because of this biogas should not be used at pressures below one column inch of water, measured by a pressure gauge.

Frequent smell checks must be performed on a biogas system. If biogas is smelled anywhere windows and doors should be opened immediately. If there is a fire the gas should be shut off at the gate valve of the biogas system.

Landfill Gas

Landfill gas is produced by wet organic waste decomposing under anaerobic conditions in a biogas.

The waste is covered and mechanically compressed by the weight of the material that is deposited above. This material prevents oxygen exposure thus allowing anaerobic microbes to thrive. This gas builds up and is slowly released into the atmosphere if the site has not been engineered to capture the gas. Landfill gas released in an uncontrolled way can be hazardous since it can become explosive when it escapes from the landfill and mixes with oxygen. The lower explosive limit is 5% methane and the upper is 15% methane.

The methane in biogas is 20 times more potent a greenhouse gas than carbon dioxide. Therefore, uncontained landfill gas, which escapes into the atmosphere may significantly contribute to the effects of global warming. In addition, volatile organic compounds (VOCs) in landfill gas contribute to the formation of photochemical smog.

Technical

Biochemical oxygen demand (BOD) is a measure of the amount of oxygen required by aerobic micro-organisms to decompose the organic matter in a sample of water. Knowing the energy density of the material being used in the biodigester as well as the BOD for the liquid discharge allows for the calculation of the daily energy output from a biodigester.

Another term related to biodigesters is effluent dirtiness, which tells how much organic material there is per unit of biogas source. Typical units for this measure are in mg BOD/litre. As an example, effluent dirtiness can range between 800–1200 mg BOD/litre in Panama.

From 1 kg of decommissioned kitchen bio-waste, 0.45 m³ of biogas can be obtained. The price for collecting biological waste from households is approximately €70 per ton.

Composition

The composition of biogas varies depending upon the origin of the anaerobic digestion process. Landfill gas typically has methane concentrations around 50%. Advanced waste treatment technologies can produce biogas with 55%–75% methane, which for reactors with free liquids can be increased to 80%-90% methane using in-situ gas purification techniques. As produced, biogas contains water vapor. The fractional volume of water vapor is a function of biogas temperature; correction of measured gas volume for water vapor content and thermal expansion is easily done via simple mathematics which yields the standardized volume of dry biogas.

Typical composition of biogas		
Compound	Formula	%
Methane	CH_4	50–75
Carbon dioxide	CO_2	25–50
Nitrogen	N_2	0–10
Hydrogen	H_2	0–1
Hydrogen sulfide	H_2S	0–3
Oxygen	O_2	0–0.5
Source: *www.kolumbus.fi, 2007*		

In some cases, biogas contains siloxanes. They are formed from the anaerobic decomposition of materials commonly found in soaps and detergents. During combustion of biogas containing siloxanes, silicon is released and can combine with free oxygen or other elements in the combustion gas. Deposits are formed containing mostly silica ($SiO2$) or silicates (Si_xO_y) and can contain calcium, sulfur, zinc, phosphorus. Such *white mineral* deposits accumulate to a surface thickness of several millimeters and must be removed by chemical or mechanical means.

Practical and cost-effective technologies to remove siloxanes and other biogas contaminants are available.

For 1000 kg (wet weight) of input to a typical biodigester, total solids may be 30% of the wet weight while volatile suspended solids may be 90% of the total solids. Protein would be 20% of the volatile solids, carbohydrates would be 70% of the volatile solids, and finally fats would be 10% of the volatile solids.

Benefits of Manure Derived Biogas

High levels of methane are produced when manure is stored under anaerobic conditions. During storage and when manure has been applied to the land, nitrous oxide is also produced as a byproduct of the denitrification process. Nitrous oxide ($N2O$) is 320 times more aggressive as a greenhouse gas than carbon dioxide and methane 25 times more than carbon dioxide.

By converting cow manure into methane biogas via anaerobic digestion, the millions of cattle in the United States would be able to produce 100 billion kilowatt hours of electricity, enough to power millions of homes across the United States. In fact, one cow can produce enough manure in one day to generate 3 kilowatt hours of electricity; only 2.4 kilowatt hours of electricity are needed to power a single 100-watt light bulb for one day. Furthermore, by converting cattle manure into methane biogas instead of letting it decompose, global warming gases could be reduced by 99 million metric tons or 4%.

Applications

Biogas can be used for electricity production on sewage works, in a CHP gas engine, where the waste heat from the engine is conveniently used for heating the digester; cooking; space heating; water heating; and process heating. If compressed, it can replace compressed natural gas for use in vehicles, where it can fuel an internal combustion engine or fuel cells and is a much more effective displacer of carbon dioxide than the normal use in on-site CHP plants.

A biogas bus in Linköping, Sweden

Biogas Upgrading

Raw biogas produced from digestion is roughly 60% methane and 29% CO_2 with trace elements of H_2S; it is not of high enough quality to be used as fuel gas for machinery. The corrosive nature of H_2S alone is enough to destroy the internals of a plant.

Methane in biogas can be concentrated via a biogas upgrader to the same standards as fossil natural gas, which itself has to go through a cleaning process, and becomes *biomethane*. If the local gas network allows, the producer of the biogas may use their distribution networks. Gas must be very clean to reach pipeline quality and must be of the correct composition for the distribution network to accept. Carbon dioxide, water, hydrogen sulfide, and particulates must be removed if present.

There are four main methods of upgrading: water washing, pressure swing adsorption, selexol adsorption, and amine gas treating. In addition to these, the use of membrane separation technology for biogas upgrading is increasing, and there are already several plants operating in Europe and USA.

The most prevalent method is water washing where high pressure gas flows into a column where the carbon dioxide and other trace elements are scrubbed by cascading water running counter-flow to the gas. This arrangement could deliver 98% methane with manufacturers guaranteeing maximum 2% methane loss in the system. It takes roughly between 3% and 6% of the total energy output in gas to run a biogas upgrading system.

Biogas Gas-grid Injection

Gas-grid injection is the injection of biogas into the methane grid (natural gas grid). Injections includes biogas until the breakthrough of micro combined heat and power two-thirds of all the energy produced by biogas power plants was lost (the heat), using the grid to transport the gas to customers, the electricity and the heat can be used for on-site generation resulting in a reduction of losses in the transportation of energy. Typical energy losses in natural gas transmission systems range from 1% to 2%. The current energy losses on a large electrical system range from 5% to 8%.

Biogas in Transport

If concentrated and compressed, it can be used in vehicle transportation. Compressed biogas is becoming widely used in Sweden, Switzerland, and Germany. A biogas-powered train, named Biogaståget Amanda (The Biogas Train Amanda), has been in service in Sweden since 2005. Biogas powers automobiles. In 1974, a British documentary film titled *Sweet as a Nut* detailed the biogas production process from pig manure and showed how it fueled a custom-adapted combustion engine. In 2007, an estimated 12,000 vehicles were being fueled with upgraded biogas worldwide, mostly in Europe.

"Biogaståget Amanda" ("The Biogas Train Amanda") train near Linköping station, Sweden

Measuring in Biogas Environments

Biogas is part of the wet gas and condensing gas (or air) category that includes mist or fog in the gas stream. The mist or fog is predominately water vapor that condenses on the sides of pipes or stacks throughout the gas flow. Biogas environments include wastewater digesters, landfills, and animal feeding operations (covered livestock lagoons).

Ultrasonic flow meters are one of the few devices capable of measuring in a biogas atmosphere. Most thermal flow meters are unable to provide reliable data because the moisture causes steady high flow readings and continuous flow spiking, although there are single-point insertion thermal mass flow meters capable of accurately monitoring biogas flows with minimal pressure drop. They can handle moisture variations that occur in the flow stream because of daily and seasonal temperature fluctuations, and account for the moisture in the flow stream to produce a dry gas value.

Legislation

European Union

The European Union has legislation regarding waste management and landfill sites called the Landfill Directive.

Countries such as the United Kingdom and Germany now have legislation in force that provides farmers with long-term revenue and energy security.

United States

The United States legislates against landfill gas as it contains VOCs. The United States Clean Air Act and Title 40 of the Code of Federal Regulations (CFR) requires landfill owners to estimate the quantity of non-methane organic compounds (NMOCs) emitted. If the estimated NMOC emissions exceeds 50 tonnes per year, the landfill owner is required to collect the gas and treat it to remove the entrained NMOCs. Treatment of the landfill gas is usually by combustion. Because of the remoteness of landfill sites, it is sometimes not economically feasible to produce electricity from the gas.

Global Developments

United States

With the many benefits of biogas, it is starting to become a popular source of energy and is starting to be used in the United States more. In 2003, the United States consumed 147 trillion BTU of energy from "landfill gas", about 0.6% of the total U.S. natural gas consumption. Methane biogas derived from cow manure is being tested in the U.S. According to a 2008 study, collected by the *Science and Children* magazine, methane biogas from cow manure would be sufficient to produce 100 billion kilowatt hours enough to power millions of homes across America. Furthermore, methane biogas has been tested to prove that it can reduce 99 million metric tons of greenhouse gas emissions or about 4% of the greenhouse gases produced by the United States.

In Vermont, for example, biogas generated on dairy farms was included in the CVPS Cow Power program. The program was originally offered by Central Vermont Public Service Corporation as a voluntary tariff and now with a recent merger with Green

Mountain Power is now the GMP Cow Power Program. Customers can elect to pay a premium on their electric bill, and that premium is passed directly to the farms in the program. In Sheldon, Vermont, Green Mountain Dairy has provided renewable energy as part of the Cow Power program. It started when the brothers who own the farm, Bill and Brian Rowell, wanted to address some of the manure management challenges faced by dairy farms, including manure odor, and nutrient availability for the crops they need to grow to feed the animals. They installed an anaerobic digester to process the cow and milking center waste from their 950 cows to produce renewable energy, a bedding to replace sawdust, and a plant-friendly fertilizer. The energy and environmental attributes are sold to the GMP Cow Power program. On average, the system run by the Rowells produces enough electricity to power 300 to 350 other homes. The generator capacity is about 300 kilowatts.

In Hereford, Texas, cow manure is being used to power an ethanol power plant. By switching to methane biogas, the ethanol power plant has saved 1000 barrels of oil a day. Over all, the power plant has reduced transportation costs and will be opening many more jobs for future power plants that will rely on biogas.

In Oakley, Kansas, an ethanol plant considered to be one of the largest biogas facilities in North America is using Integrated Manure Utilization System "IMUS" to produce heat for its boilers by utilizing feedlot manure, municipal organics and ethanol plant waste. At full capacity the plant is expected to replace 90% of the fossil fuel used in the manufacturing process of ethanol.

Europe

The level of development varies greatly in Europe. While countries such as Germany, Austria and Sweden are fairly advanced in their use of biogas, there is a vast potential for this renewable energy source in the rest of the continent, especially in Eastern Europe. Different legal frameworks, education schemes and the availability of technology are among the prime reasons behind this untapped potential. Another challenge for the further progression of biogas has been negative public perception.

Initiated by the events of the gas crisis in Europe during December 2008, it was decided to launch the EU project "SEBE" (Sustainable and Innovative European Biogas Environment) which is financed under the CENTRAL programme. The goal is to address the energy dependence of Europe by establishing an online platform to combine knowledge and launch pilot projects aimed at raising awareness among the public and developing new biogas technologies.

In February 2009, the European Biogas Association (EBA) was founded in Brussels as a non-profit organisation to promote the deployment of sustainable biogas production and use in Europe. EBA's strategy defines three priorities: establish biogas as an important part of Europe's energy mix, promote source separation of household waste to increase the gas potential, and support the production of biomethane as vehicle fuel. In July 2013, it had 60 members from 24 countries across Europe.

UK

As of September 2013, there are about 130 non-sewage biogas plants in the UK. Most are on-farm, and some larger facilities exist off-farm, which are taking food and consumer wastes.

On 5 October 2010, biogas was injected into the UK gas grid for the first time. Sewage from over 30,000 Oxfordshire homes is sent to Didcot sewage treatment works, where it is treated in an anaerobic digestor to produce biogas, which is then cleaned to provide gas for approximately 200 homes.

In 2015 the Green-Energy company Ecotricity announced their plans to build three grid-injecting digester's.

Germany

Germany is Europe's biggest biogas producer and the market leader in biogas technology. In 2010 there were 5,905 biogas plants operating throughout the country: Lower Saxony, Bavaria, and the eastern federal states are the main regions. Most of these plants are employed as power plants. Usually the biogas plants are directly connected with a CHP which produces electric power by burning the bio methane. The electrical power is then fed into the public power grid. In 2010, the total installed electrical capacity of these power plants was 2,291 MW. The electricity supply was approximately 12.8 TWh, which is 12.6% of the total generated renewable electricity.

Biogas in Germany is primarily extracted by the co-fermentation of energy crops (called 'NawaRo', an abbreviation of *nachwachsende Rohstoffe*, German for renewable resources) mixed with manure. The main crop used is corn. Organic waste and industrial and agricultural residues such as waste from the food industry are also used for biogas generation.In this respect, biogas production in Germany differs significantly from the UK, where biogas generated from landfill sites is most common.

Biogas production in Germany has developed rapidly over the last 20 years. The main reason is the legally created frameworks. Government support of renewable energy started in 1991 with the Electricity Feed-in Act (*StrEG*). This law guaranteed the producers of energy from renewable sources the feed into the public power grid, thus the power companies were forced to take all produced energy from independent private producers of green energy. In 2000 the Electricity Feed-in Act was replaced by the Renewable Energy Sources Act (*EEG*). This law even guaranteed a fixed compensation for the produced electric power over 20 years. The amount of around 8 ¢/kWh gave farmers the opportunity to become energy suppliers and gain a further source of income.

The German agricultural biogas production was given a further push in 2004 by implementing the so-called NawaRo-Bonus. This is a special payment given for the use of renewable resources, that is, energy crops. In 2007 the German government stressed its intention to invest further effort and support in improving the renewable energy supply to provide an answer on growing climate challenges and increasing oil prices by the 'Integrated Climate and Energy Programme'.

This continual trend of renewable energy promotion induces a number of challenges facing the management and organisation of renewable energy supply that has also several impacts on the biogas production. The first challenge to be noticed is the high area-consuming of the biogas electric power supply. In 2011 energy crops for biogas production consumed an area of circa 800,000 ha in Germany. This high demand of agricultural areas generates new competitions with the food industries that did not exist hitherto. Moreover, new industries and markets were created in predominately rural regions entailing different new players with an economic, political and civil background. Their influence and acting has to be governed to gain all advantages this new source of energy is offering. Finally biogas will furthermore play an important role in the German renewable energy supply if good governance is focused.

Indian Subcontinent

Biogas in India has been traditionally based on dairy manure as feed stock and these "gobar" gas plants have been in operation for a long period of time, especially in rural India. In the last 2-3 decades, research organisations with a focus on rural energy security have enhanced the design of the systems resulting in newer efficient low cost designs such as the Deenabandhu model.

The Deenabandhu Model is a new biogas-production model popular in India. (*Deenabandhu* means "friend of the helpless.") The unit usually has a capacity of 2 to 3 cubic metres. It is constructed using bricks or by a ferrocement mixture. In India, the brick model costs slightly more than the ferrocement model; however, India's Ministry of New and Renewable Energy offers some subsidy per model constructed.

LPG (Liquefied Petroleum Gas) is a key source of cooking fuel in urban India and its prices have been increasing along with the global fuel prices. Also the heavy subsidies provided by the successive governments in promoting LPG as a domestic cooking fuel has become a financial burden renewing the focus on biogas as a cooking fuel alternative in urban establishments. This has led to the development of prefabricated digester for modular deployments as compared to RCC and cement structures which take a longer duration to construct. Renewed focus on process technology like the Biourja process model has enhanced the stature of medium and large scale anaerobic digester in India as a potential alternative to LPG as primary cooking fuel.

In India, Nepal, Pakistan and Bangladesh biogas produced from the anaerobic digestion of manure in small-scale digestion facilities is called gobar gas; it is estimated that such facilities exist in over 2 million households in India, 50,000 in Bangladesh and thousands in Pakistan, particularly North Punjab, due to the thriving population of livestock. The digester is an airtight circular pit made of concrete with a pipe connection. The manure is directed to the pit, usually straight from the cattle shed. The pit is filled with a required quantity of wastewater. The gas pipe is connected to the kitchen fireplace through control valves. The combustion of this biogas has very little odour or smoke. Owing to simplicity in implementation and use of cheap raw materials in villages, it is one of the most environmentally sound energy sources for rural needs. One type of these system is the Sintex Digester. Some designs use vermiculture to further enhance the slurry produced by the biogas plant for use as compost.

To create awareness and associate the people interested in biogas, the Indian Biogas Association was formed. It aspires to be a unique blend of nationwide operators, manufacturers and planners of biogas plants, and representatives from science and research. The association was founded in 2010 and is now ready to start mushrooming. Its motto is "propagating Biogas in a sustainable way".

In Pakistan, the Rural Support Programmes Network is running the Pakistan Domestic Biogas Programme which has installed 5,360 biogas plants and has trained in excess of 200 masons on the technology and aims to develop the Biogas Sector in Pakistan.

In Nepal, the government provides subsidies to build biogas plant.

China

The Chinese had experimented the applications of biogas since 1958. Around 1970, China had installed 6,000,000 digesters in an effort to make agriculture more efficient. During the last years the technology has met high growth rates. This seems to be the earliest developments in generating biogas from agricultural waste.

In Developing Nations

Domestic biogas plants convert livestock manure and night soil into biogas and slurry, the fermented manure. This technology is feasible for small-holders with livestock producing 50 kg manure per day, an equivalent of about 6 pigs or 3 cows. This manure has to be collectable to mix it with water and feed it into the plant. Toilets can be connected. Another precondition is the temperature that affects the fermentation process. With an optimum at 36 C° the technology especially applies for those living in a (sub) tropical climate. This makes the technology for small holders in developing countries often suitable.

Depending on size and location, a typical brick made fixed dome biogas plant can be installed at the yard of a rural household with the investment between US$300 to $500 in Asian countries and up to $1400 in the African context. A high quality biogas plant needs minimum maintenance costs and can produce gas for at least 15–20 years without major problems and re-investments. For the user, biogas provides clean cooking energy, reduces indoor air pollution, and reduces the time needed for traditional biomass collection, especially for women and children. The slurry is a clean organic fertilizer that potentially increases agricultural productivity.

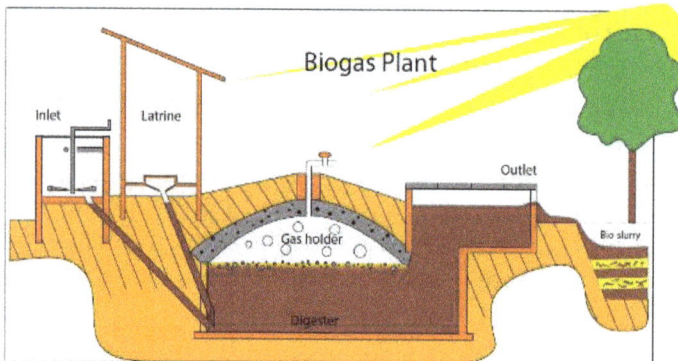

Simple sketch of household biogas plant

Domestic biogas technology is a proven and established technology in many parts of the world, especially Asia. Several countries in this region have embarked on large-scale programmes on domestic biogas, such as China and India.

The Netherlands Development Organisation, SNV, supports national programmes on domestic biogas that aim to establish commercial-viable domestic biogas

sectors in which local companies market, install and service biogas plants for households. In Asia, SNV is working in Nepal, Vietnam, Bangladesh, Bhutan, Cambodia, Lao PDR, Pakistan and Indonesia, and in Africa; Rwanda, Senegal, Burkina Faso, Ethiopia, Tanzania, Uganda, Kenya, Benin and Cameroon.

In South Africa a prebuilt Biogas system is manufactured and sold. One key feature is that installation requires less skill and is quicker to install as the digester tank is premade plastic.

Society and Culture

In the 1985 Australian film *Mad Max Beyond Thunderdome* the post-apocalyptic settlement Barter town is powered by a central biogas system based upon a piggery. As well as providing electricity, methane is used to power Barter's vehicles.

"Cow Town", written in the early 1940s, discuss the travails of a city vastly built on cow manure and the hardships brought upon by the resulting methane biogas. Carter McCormick, an engineer from a town outside the city, is sent in to figure out a way to utilize this gas to help power, rather than suffocate, the city.

Biofuel

A biofuel is a fuel that is produced through contemporary biological processes, such as agriculture and anaerobic digestion, rather than a fuel produced by geological process-es such as those involved in the formation of fossil fuels, such as coal and petroleum, from prehistoric biological matter. Biofuels can be derived directly from plants, or in-directly from agricultural, commercial, domestic, and/or industrial wastes. Renewable biofuels generally involve contemporary carbon fixation, such as those that occur in plants or microalgae through the process of photosynthesis. Other renewable biofuels are made through the use or conversion of biomass (referring to recently living organ-isms, most often referring to plants or plant-derived materials). This biomass can be converted to convenient energy-containing substances in three different ways: thermal conversion, chemical conversion, and biochemical conversion. This biomass conver-sion can result in fuel in solid, liquid, or gas form. This new biomass can also be used directly for biofuels.

A bus fueled by biodiesel

Information on pump regarding ethanol fuel blend up to 10%, California

Bioethanol is an alcohol made by fermentation, mostly from carbohydrates produced in sugar or starch crops such as corn, sugarcane, or sweet sorghum. Cellulosic biomass, derived from non-food sources, such as trees and grasses, is also being developed as a feedstock for ethanol production. Ethanol can be used as a fuel for vehicles in its pure form, but it is usually used as a gasoline additive to increase octane and improve vehicle emissions. Bioethanol is widely used in the USA and in Brazil. Current plant design does not provide for converting the lignin portion of plant raw materials to fuel components by fermentation.

Biodiesel can be used as a fuel for vehicles in its pure form, but it is usually used as a diesel additive to reduce levels of particulates, carbon monoxide, and hydrocarbons from diesel-powered vehicles. Biodiesel is produced from oils or fats using transesterification and is the most common biofuel in Europe.

In 2010, worldwide biofuel production reached 105 billion liters (28 billion gallons US), up 17% from 2009, and biofuels provided 2.7% of the world's fuels for road transport. Global ethanol fuel production reached 86 billion liters (23 billion gallons US) in 2010, with the United States and Brazil as the world's top producers, accounting together for 90% of global production. The world's largest biodiesel producer is the European Union, accounting for 53% of all biodiesel production in 2010. As of 2011, mandates for blending biofuels exist in 31 countries at the national level and in 29 states or provinces. The International Energy Agency has a goal for biofuels to meet more than a quarter of world demand for transportation fuels by 2050 to reduce dependence on petroleum and coal. The production of biofuels also led into a flourishing automotive industry, where by 2010, 79% of all cars produced in Brazil were made with a hybrid fuel system of bioethanol and gasoline.

There are various social, economic, environmental and technical issues relating to biofuels production and use, which have been debated in the popular media and scientific journals. These include: the effect of moderating oil prices, the "food vs fuel" debate, poverty reduction potential, carbon emissions levels, sustainable biofuel production, deforestation and soil erosion, loss of biodiversity, impact on water resources, rural social exclusion and injustice, shantytown migration, rural unskilled unemployment, and nitrous oxide (NO_2) emissions.

Liquid Fuels for Transportation

Most transportation fuels are liquids, because vehicles usually require high energy density. This occurs naturally in liquids and solids. High energy density can also be provided by an internal combustion engine. These engines require clean-burning fuels. The fuels that are easiest to burn cleanly are typically liquids and gases. Thus, liquids meet the requirements of being both energy-dense and clean-burning. In addition, liquids (and gases) can be pumped, which means handling is easily mechanized, and thus less laborious.

First-generation Biofuels

"First-generation" or conventional biofuels are made from sugar, starch, or vegetable oil.

Ethanol

Biologically produced alcohols, most commonly ethanol, and less commonly propanol and butanol, are produced by the action of microorganisms and enzymes through the fermentation of sugars or starches (easiest), or cellulose (which is more difficult). Biobutanol (also called biogasoline) is often claimed to provide a direct replacement for gasoline, because it can be used directly in a gasoline engine.

Neat ethanol on the left (A), gasoline on the right (G) at a filling station in Brazil

Ethanol fuel is the most common biofuel worldwide, particularly in Brazil. Alcohol fuels are produced by fermentation of sugars derived from wheat, corn, sugar beets, sugar cane, molasses and any sugar or starch from which alcoholic beverages such as whiskey, can be made (such as potato and fruit waste, etc.). The ethanol production methods used are enzyme digestion (to release sugars from stored starches), fermentation of the sugars, distillation and drying. The distillation process requires significant energy input for heat (sometimes unsustainable natural gas fossil fuel, but cellulosic biomass such as bagasse, the waste left after sugar cane is pressed to extract its juice, is the most common fuel in Brazil, while pellets, wood chips and also waste heat are more common in Europe) Waste steam fuels ethanol factory- where waste heat from the factories also is used in the district heating grid.

U.S. President George W. Bush looks at sugar cane, a source of biofuel, with Brazilian President Luiz Inácio Lula da Silva during a tour on biofuel technology at Petrobras in São Paulo, Brazil, 9 March 2007.

Ethanol can be used in petrol engines as a replacement for gasoline; it can be mixed with gasoline to any percentage. Most existing car petrol engines can run on blends of up to 15% bioethanol with petroleum/gasoline. Ethanol has a smaller energy density than that of gasoline; this means it takes more fuel (volume and mass) to produce the same amount of work. An advantage of ethanol (CH_3CH_2OH) is that it has a higher octane rating than ethanol-free gasoline available at roadside gas stations, which allows an increase of an engine's compression ratio for increased thermal efficiency. In high-altitude (thin air) locations, some states mandate a mix of gasoline and ethanol as a winter oxidizer to reduce atmospheric pollution emissions.

Ethanol is also used to fuel bioethanol fireplaces. As they do not require a chimney and are "flueless", bioethanol fires are extremely useful for newly built homes and apartments without a flue. The downsides to these fireplaces is that their heat output is slightly less than electric heat or gas fires, and precautions must be taken to avoid carbon monoxide poisoning.

Corn-to-ethanol and other food stocks has led to the development of cellulosic ethanol. According to a joint research agenda conducted through the US Department of Energy, the fossil energy ratios (FER) for cellulosic ethanol, corn ethanol, and gasoline are 10.3, 1.36, and 0.81, respectively.

Ethanol has roughly one-third lower energy content per unit of volume compared to gasoline. This is partly counteracted by the better efficiency when using ethanol (in a long-term test of more than 2.1 million km, the BEST project found FFV vehicles to be 1-26 % more energy efficient than petrol cars The BEST project), but the volumetric consumption increases by approximately 30%, so more fuel stops are required.

With current subsidies, ethanol fuel is slightly cheaper per distance traveled in the United States.

Biodiesel

Biodiesel is the most common biofuel in Europe. It is produced from oils or fats using transesterification and is a liquid similar in composition to fossil/mineral diesel. Chemically, it consists mostly of fatty acid methyl (or ethyl) esters (FAMEs). Feedstocks for biodiesel include animal fats, vegetable oils, soy, rapeseed, jatropha, mahua, mustard, flax, sunflower, palm oil, hemp, field pennycress, *Pongamia pinnata* and algae. Pure biodiesel (B100) currently reduces emissions with up to 60% compared to diesel Second generation B100.

Biodiesel can be used in any diesel engine when mixed with mineral diesel. In some countries, manufacturers cover their diesel engines under warranty for B100 use, although Volkswagen of Germany, for example, asks drivers to check by telephone with the VW environmental services department before switching to B100. B100 may become more viscous at lower temperatures, depending on the feedstock used. In most cases, biodiesel is compatible with diesel engines from 1994 onwards, which use 'Viton' (by DuPont) synthetic rubber in their mechanical fuel injection systems. Note however, that no vehicles are certified for using neat biodiesel before 2014, as there was no emission control protocol available for biodiesel before this date.

Electronically controlled 'common rail' and 'unit injector' type systems from the late 1990s onwards may only use biodiesel blended with conventional diesel fuel. These engines have finely metered and atomized multiple-stage injection systems that are very sensitive to the viscosity of the fuel. Many current-generation diesel engines are made so that they can run on B100 without altering the engine itself, although this depends on the fuel rail design. Since biodiesel is an effective solvent and cleans residues deposited by mineral diesel, engine filters may need to be replaced more often, as the biofuel dissolves old deposits in the fuel tank and pipes. It also effectively cleans the engine combustion chamber of carbon deposits, helping to maintain efficiency. In many European countries, a 5% biodiesel blend is widely used and is available at thousands of gas stations. Biodiesel is also an oxygenated fuel, meaning it contains a reduced amount of carbon and higher hydrogen and oxygen content than fossil diesel. This improves the combustion of biodiesel and reduces the particulate emissions from unburnt carbon. However, using neat biodiesel may increase NOx-emissions Nylund.N-O & Koponen.K. 2013. Fuel and Technology Alternatives for Buses. Overall Energy Efficiency and Emission Performance. IEA Bioenergy Task 46. Possibly the new emission standards Euro VI/EPA 10 will lead to reduced NOx-levels also when using B100.

Biodiesel is also safe to handle and transport because it is non-toxic and biodegradable, and has a high flash point of about 300 °F (148 °C) compared to petroleum diesel fuel, which has a flash point of 125 °F (52 °C).

In the USA, more than 80% of commercial trucks and city buses run on diesel. The emerging US biodiesel market is estimated to have grown 200% from 2004 to 2005. "By the end of 2006 biodiesel production was estimated to increase fourfold [from 2004] to more than" 1 billion US gallons (3,800,000 m³).

In France, biodiesel is incorporated at a rate of 8% in the fuel used by all French diesel vehicles. Avril Group produces under the brand Diester, a fifth of 11 million tons of biodiesel consumed annually by the European Union. It is the leading European producer of biodiesel.

Other Bioalcohols

Methanol is currently produced from natural gas, a non-renewable fossil fuel. In the future it is hoped to be produced from biomass as biomethanol. This is technically feasible, but the production is currently being postponed for concerns of Jacob S. Gibbs and Brinsley Coleberd that the economic viability is still pending. The methanol economy is an alternative to the hydrogen economy, compared to today's hydrogen production from natural gas.

Butanol (C4H9OH) is formed by ABE fermentation (acetone, butanol, ethanol) and experimental modifications of the process show potentially high net energy gains with butanol as the only liquid product. Butanol will produce more energy and allegedly can be burned "straight" in existing gasoline engines (without modification to the engine or car), and is less corrosive and less water-soluble than ethanol, and could be distributed via existing infrastructures. DuPont and BP are working together to help develop butanol. *E. coli* strains have also been successfully engineered to produce butanol by modifying their amino acid metabolism.

Green Diesel

Green diesel is produced through hydrocracking biological oil feedstocks, such as vegetable oils and animal fats. Hydrocracking is a refinery method that uses elevated temperatures and pressure in the presence of a catalyst to break down larger molecules, such as those found in vegetable oils, into shorter hydrocarbon chains used in diesel engines. It may also be called renewable diesel, hydrotreated vegetable oil or hydrogen-derived renewable diesel. Green diesel has the same chemical properties as petroleum-based diesel. It does not require new engines, pipelines or infrastructure to distribute and use, but has not been produced at a cost that is competitive with petroleum. Gasoline versions are also being developed. Green diesel is being developed in Louisiana and Singapore by ConocoPhillips, Neste Oil, Valero, Dynamic Fuels, and Honeywell UOP as well as Preem in Gothenburg, Sweden, creating what is known as Evolution Diesel.

Biofuel Gasoline

In 2013 UK researchers developed a genetically modified strain of Escherichia coli (E.Coli), which could transform glucose into biofuel gasoline that does not need to be blended. Later in 2013 UCLA researchers engineered a new metabolic pathway to bypass glycolysis and increase the rate of conversion of sugars into biofuel, while KAIST researchers developed a strain capable of producing short-chain alkanes, free fatty acids, fatty esters and fatty alcohols through the fatty acyl (acyl carrier protein (ACP)) to fatty acid to fatty acyl-CoA pathway *in vivo*. It is believed that in the future it will be possible to "tweak" the genes to make gasoline from straw or animal manure.

Vegetable Oil

Straight unmodified edible vegetable oil is generally not used as fuel, but lower-quality oil can and has been used for this purpose. Used vegetable oil is increasingly being processed into biodiesel, or (more rarely) cleaned of water and particulates and used as a fuel.

As with 100% biodiesel (B100), to ensure the fuel injectors atomize the vegetable oil in the correct pattern for efficient combustion, vegetable oil fuel must be heated to reduce its viscosity to that of diesel, either by electric coils or heat exchangers. This is easier in warm or temperate climates. MAN B&W Diesel, Wärtsilä, and Deutz AG, as well as a number of smaller companies, such as Elsbett, offer engines that are compatible with straight vegetable oil, without the need for after-market modifications.

Filtered waste vegetable oil

Vegetable oil can also be used in many older diesel engines that do not use common rail or unit injection electronic diesel injection systems. Due to the design of the combustion chambers in indirect injection engines, these are the best engines for use with vegetable oil. This system allows the relatively larger oil molecules more time to burn. Some older engines, especially Mercedes, are driven experimentally by enthusiasts without any conversion, a handful of drivers have experienced limited success with earlier pre-"Pumpe Duse" VW TDI engines and other similar engines with direct injection. Several companies, such as Elsbett or Wolf, have developed professional conversion kits and successfully installed hundreds of them over the last decades.

Walmart's truck fleet logs millions of miles each year, and the company planned to double the fleet's efficiency between 2005 and 2015. This truck is one of 15 based at Walmart's Buckeye, Arizona distribution center that was converted to run on a biofuel made from reclaimed cooking grease produced during food preparation at Walmart stores.

Oils and fats can be hydrogenated to give a diesel substitute. The resulting product is a straight-chain hydrocarbon with a high cetane number, low in aromatics and sulfur and does not contain oxygen. Hydrogenated oils can be blended with diesel in all proportions. They have several advantages over biodiesel, including good performance at low temperatures, no storage stability problems and no susceptibility to microbial attack.

Bioethers

Bioethers (also referred to as fuel ethers or oxygenated fuels) are cost-effective compounds that act as octane rating enhancers."Bioethers are produced by the reaction of reactive iso-olefins, such as iso-butylene, with bioethanol." Bioethers are created by wheat or sugar beet. They also enhance engine performance, whilst significantly reducing engine wear and toxic exhaust emissions. Though bioethers are likely to replace petroethers in the UK, it is highly unlikely they will become a fuel in and of itself due to the low energy density. Greatly reducing the amount of ground-level ozone emissions, they contribute to air quality.

When it comes to transportation fuel there are six ether additives- 1. Dimethyl Ether (DME) 2. Diethyl Ether (DEE) 3. Methyl Teritiary-Butyl Ether (MTBE) 4. Ethyl *ter*-butyl ether (ETBE) 5. *Ter*-amyl methyl ether (TAME) 6. *Ter*-amyl ethyl Ether (TAEE)

The European Fuel Oxygenates Association (aka EFOA) credits Methyl Tertiary-Butyl Ether (MTBE) and Ethyl ter-butyl ether (ETBE) as the most commonly used ethers in fuel to replace lead. Ethers were brought into fuels in Europe in the 1970s to replace the highly toxic compound. Although Europeans still use Bio-ether additives, the US no longer has an oxygenate requirement therefore bio-ethers are no longer used as the main fuel additive.

Biogas

Biogas is methane produced by the process of anaerobic digestion of organic material by anaerobes. It can be produced either from biodegradable waste materials or by the use of energy crops fed into anaerobic digesters to supplement gas yields. The solid byproduct, digestate, can be used as a biofuel or a fertilizer.

Pipes carrying biogas

- Biogas can be recovered from mechanical biological treatment waste processing systems.

  Note: Landfill gas, a less clean form of biogas, is produced in landfills through naturally occurring anaerobic digestion. If it escapes into the atmosphere, it is a potential greenhouse gas.

- Farmers can produce biogas from manure from their cattle by using anaerobic digesters.

Syngas

Syngas, a mixture of carbon monoxide, hydrogen and other hydrocarbons, is produced by partial combustion of biomass, that is, combustion with an amount of oxygen that is not sufficient to convert the biomass completely to carbon dioxide and water. Before partial combustion, the biomass is dried, and sometimes pyrolysed. The resulting gas mixture, syngas, is more efficient than direct combustion of the original biofuel; more of the energy contained in the fuel is extracted.

- Syngas may be burned directly in internal combustion engines, turbines or high-temperature fuel cells. The wood gas generator, a wood-fueled gasification reactor, can be connected to an internal combustion engine.

- Syngas can be used to produce methanol, DME and hydrogen, or converted via the Fischer-Tropsch process to produce a diesel substitute, or a mixture of alcohols that can be blended into gasoline. Gasification normally relies on temperatures greater than 700 °C.

- Lower-temperature gasification is desirable when co-producing biochar, but results in syngas polluted with tar.

Solid Biofuels

Examples include wood, sawdust, grass trimmings, domestic refuse, charcoal, agricultural waste, nonfood energy crops, and dried manure.

When raw biomass is already in a suitable form (such as firewood), it can burn directly in a stove or furnace to provide heat or raise steam. When raw biomass is in an inconvenient form (such as sawdust, wood chips, grass, urban waste wood, agricultural residues), the typical process is to densify the biomass. This process includes grinding the raw biomass to an appropriate particulate size (known as hogfuel), which, depending on the densification type, can be from 1 to 3 cm (0.4 to 1.2 in), which is then concentrated into a fuel product. The current processes produce wood pellets, cubes, or pucks. The pellet process is most common in Europe, and is typically a pure wood product. The other types of densification are larger in size compared to a pellet, and are compatible with a broad range of input feedstocks. The resulting densified fuel is easier to transport and feed into thermal generation systems, such as boilers.

Industry has used sawdust, bark and chips for fuel for decades, primary in the pulp and paper industry, and also bagasse (spent sugar cane) fueled boilers in the sugar cane industry. Boilers in the range of 500,000 lb/hr of steam, and larger, are in routine operation, using grate, spreader stoker, suspension burning and fluid bed combustion. Utilities generate power, typically in the range of 5 to 50 MW, using locally available fuel. Other industries have also installed wood waste fueled boilers and dryers in areas with low cost fuel.

One of the advantages of biomass fuel is that it is often a byproduct, residue or waste-product of other processes, such as farming, animal husbandry and forestry. In theory, this means fuel and food production do not compete for resources, although this is not always the case.

A problem with the combustion of raw biomass is that it emits considerable amounts of pollutants, such as particulates and polycyclic aromatic hydrocarbons. Even modern pellet boilers generate much more pollutants than oil or natural gas boilers. Pellets made from agricultural residues are usually worse than wood pellets, producing much larger emissions of dioxins and chlorophenols.

In spite of the above noted study, numerous studies have shown biomass fuels have significantly less impact on the environment than fossil based fuels. Of note is the US Department of Energy Laboratory, operated by Midwest Research Institute Biomass Power and Conventional Fossil Systems with and without CO2 Sequestration – Comparing the Energy Balance, Greenhouse Gas Emissions and Economics Study. Power generation emits significant amounts of greenhouse gases (GHGs), mainly carbon dioxide (CO_2). Sequestering CO_2 from the power plant flue gas can significantly reduce the GHGs from the power plant itself, but this is not the total picture. CO_2 capture and sequestration consumes additional energy, thus lowering the plant's fuel-to-electricity efficiency. To compensate for this, more fossil fuel must be procured and consumed to make up for lost capacity.

Taking this into consideration, the global warming potential (GWP), which is a combination of CO_2, methane (CH_4), and nitrous oxide (N_2O) emissions, and energy balance of the system need to be examined using a life cycle assessment. This takes into account the upstream processes which remain constant after CO_2 sequestration, as well as the steps required for additional power generation. Firing biomass instead of coal led to a 148% reduction in GWP.

A derivative of solid biofuel is biochar, which is produced by biomass pyrolysis. Biochar made from agricultural waste can substitute for wood charcoal. As wood stock becomes scarce, this alternative is gaining ground. In eastern Democratic Republic of Congo, for example, biomass briquettes are being marketed as an alternative to charcoal to protect Virunga National Park from deforestation associated with charcoal production.

Second-generation (Advanced) Biofuels

Second generation biofuels, also known as advanced biofuels, are fuels that can be manufactured from various types of biomass. Biomass is a wide-ranging term meaning any source of organic carbon that is renewed rapidly as part of the carbon cycle. Biomass is derived from plant materials but can also include animal materials.

First generation biofuels are made from the sugars and vegetable oils found in arable crops, which can be easily extracted using conventional technology. In comparison, second generation biofuels are made from lignocellulosic biomass or woody crops, agricultural residues or waste, which makes it harder to extract the required fuel. A series of physical and chemical treatments might be required to convert lignocellulosic biomass to liquid fuels suitable for transportation.

Sustainable Biofuels

Biofuels in the form of liquid fuels derived from plant materials, are entering the market, driven mainly by the perception that they reduce climate gas emissions, and also by factors such as oil price spikes and the need for increased energy security. However, many of the biofuels that are currently being supplied have been criticised for their adverse impacts on the natural environment, food security, and land use. In 2008, the Nobel-prize winning chemist Paul J. Crutzen published findings that the release of nitrous oxide (N_2O) emissions in the production of biofuels means that overall they contribute more to global warming than the fossil fuels they replace.

The challenge is to support biofuel development, including the development of new cellulosic technologies, with responsible policies and economic instruments to help ensure that biofuel commercialization is sustainable. Responsible commercialization of biofuels represents an opportunity to enhance sustainable economic prospects in Africa, Latin America and Asia.

According to the Rocky Mountain Institute, sound biofuel production practices would not hamper food and fibre production, nor cause water or environmental problems, and would enhance soil fertility. The selection of land on which to grow the feedstocks is a critical component of the ability of biofuels to deliver sustainable solutions. A key consideration is the minimisation of biofuel competition for prime cropland.

Biofuels by Region

There are international organizations such as IEA Bioenergy, established in 1978 by the OECD International Energy Agency (IEA), with the aim of improving cooperation and information exchange between countries that have national programs in bioenergy research, development and deployment. The UN International Biofuels Forum is formed by Brazil, China, India, Pakistan, South Africa, the United States and the European Commission. The world leaders in biofuel development and use are Brazil, the United

States, France, Sweden and Germany. Russia also has 22% of world's forest, and is a big biomass (solid biofuels) supplier. In 2010, Russian pulp and paper maker, Vyborgskaya Cellulose, said they would be producing pellets that can be used in heat and electricity generation from its plant in Vyborg by the end of the year. The plant will eventually produce about 900,000 tons of pellets per year, making it the largest in the world once operational.

Bio Diesel Powered Fast Attack Craft Of Indian Navy patrolling during IFR 2016. The green bands on the vessels are indicative of the fact that the vessels are powered by bio-diesel

Biofuels currently make up 3.1% of the total road transport fuel in the UK or 1,440 million litres. By 2020, 10% of the energy used in UK road and rail transport must come from renewable sources – this is the equivalent of replacing 4.3 million tonnes of fossil oil each year. Conventional biofuels are likely to produce between 3.7 and 6.6% of the energy needed in road and rail transport, while advanced biofuels could meet up to 4.3% of the UK's renewable transport fuel target by 2020.

Air Pollution

Biofuels are different from fossil fuels in regard to greenhouse gases but are similar to fossil fuels in that biofuels contribute to air pollution. Burning produces airborne carbon particulates, carbon monoxide and nitrous oxides. The WHO estimates 3.7 million premature deaths worldwide in 2012 due to air pollution. Brazil burns significant amounts of ethanol biofuel. Gas chromatograph studies were performed of ambient air in São Paulo, Brazil, and compared to Osaka, Japan, which does not burn ethanol fuel. Atmospheric Formaldehyde was 160% higher in Brazil, and Acetaldehyde was 260% higher.

Debates Regarding The Production and Use of Biofuel

There are various social, economic, environmental and technical issues with biofuel production and use, which have been discussed in the popular media and scientific journals. These include: the effect of moderating oil prices, the "food vs fuel" debate, food prices, poverty reduction potential, energy ratio, energy requirements, carbon emissions levels, sustainable biofuel production, deforestation and soil erosion, loss of biodiversity, impact on water resources, the possible modifications necessary to run the engine on biofuel, as well as energy balance and efficiency. The International Resource Panel, which provides independent scientific assessments and expert advice on a variety of resource-related themes, assessed the issues relating to biofuel use in its first report *Towards sustainable production and use of resources: Assessing Biofuels.* "Assessing Biofuels" outlined the wider and interrelated factors that need to be considered when deciding on the relative merits of pursuing one biofuel over another. It concluded that not all biofuels perform equally in terms of their impact on climate, energy security and ecosystems, and suggested that environmental and social impacts need to be assessed throughout the entire life-cycle.

Another issue with biofuel use and production is the US has changed mandates many times because the production has been taking longer than expected. The Renewable Fuel Standard (RFS) set by congress for 2010 was pushed back to at best 2012 to produce 100 million gallons of pure ethanol (not blended with a fossil fuel).

Current Research

Research is ongoing into finding more suitable biofuel crops and improving the oil yields of these crops. Using the current yields, vast amounts of land and fresh water would be needed to produce enough oil to completely replace fossil fuel usage. It would require twice the land area of the US to be devoted to soybean production, or two-thirds to be devoted to rapeseed production, to meet current US heating and transportation needs.

Specially bred mustard varieties can produce reasonably high oil yields and are very useful in crop rotation with cereals, and have the added benefit that the meal left over after the oil has been pressed out can act as an effective and biodegradable pesticide.

The NFESC, with Santa Barbara-based Biodiesel Industries, is working to develop biofuels technologies for the US navy and military, one of the largest diesel fuel users in the world. A group of Spanish developers working for a company called Ecofasa announced a new biofuel made from trash. The fuel is created from general urban waste which is treated by bacteria to produce fatty acids, which can be used to make biofuels.

Ethanol Biofuels

As the primary source of biofuels in North America, many organizations are conducting research in the area of ethanol production. The National Corn-to-Ethanol Research Center (NCERC) is a research division of Southern Illinois University Edwardsville dedicated solely to ethanol-based biofuel research projects. On the federal level, the USDA conducts a large amount of research regarding ethanol production in the United

States. Much of this research is targeted toward the effect of ethanol production on domestic food markets. A division of the U.S. Department of Energy, the National Renewable Energy Laboratory (NREL), has also conducted various ethanol research projects, mainly in the area of cellulosic ethanol.

Cellulosic ethanol commercialization is the process of building an industry out of methods of turning cellulose-containing organic matter into fuel. Companies, such as Iogen, POET, and Abengoa, are building refineries that can process biomass and turn it into bioethanol. Companies, such as Diversa, Novozymes, and Dyadic, are producing enzymes that could enable a cellulosic ethanol future. The shift from food crop feedstocks to waste residues and native grasses offers significant opportunities for a range of players, from farmers to biotechnology firms, and from project developers to investors.

As of 2013, the first commercial-scale plants to produce cellulosic biofuels have begun operating. Multiple pathways for the conversion of different biofuel feedstocks are being used. In the next few years, the cost data of these technologies operating at commercial scale, and their relative performance, will become available. Lessons learnt will lower the costs of the industrial processes involved.

In parts of Asia and Africa where drylands prevail, sweet sorghum is being investigated as a potential source of food, feed and fuel combined. The crop is particularly suitable for growing in arid conditions, as it only extracts one seventh of the water used by sugarcane. In India, and other places, sweet sorghum stalks are used to produce biofuel by squeezing the juice and then fermenting into ethanol.

A study by researchers at the International Crops Research Institute for the Semi-Arid Tropics (ICRISAT) found that growing sweet sorghum instead of grain sorghum could increase farmers incomes by US$40 per hectare per crop because it can provide fuel in addition to food and animal feed. With grain sorghum currently grown on over 11 million hectares (ha) in Asia and on 23.4 million ha in Africa, a switch to sweet sorghum could have a considerable economic impact.

Algae Biofuels

From 1978 to 1996, the US NREL experimented with using algae as a biofuels source in the "Aquatic Species Program". A self-published article by Michael Briggs, at the UNH Biofuels Group, offers estimates for the realistic replacement of all vehicular fuel with biofuels by using algae that have a natural oil content greater than 50%, which Briggs suggests can be grown on algae ponds at wastewater treatment plants. This oil-rich algae can then be extracted from the system and processed into biofuels, with the dried remainder further reprocessed to create ethanol. The production of algae to harvest oil for biofuels has not yet been undertaken on a commercial scale, but feasibility studies have been conducted to arrive at the above yield estimate. In addition to its projected high yield, algaculture — unlike crop-based biofuels — does not entail a decrease in food production, since it requires neither farmland nor fresh water. Many companies are pursuing algae bioreactors for various purposes, including scaling up biofuels production to commercial levels. Prof. Rodrigo E. Teixeira from the University of Alabama in Huntsville demonstrated the extraction of biofuels lipids from wet algae using a simple and economical reaction in ionic liquids.

Jatropha

Several groups in various sectors are conducting research on *Jatropha curcas*, a poisonous shrub-like tree that produces seeds considered by many to be a viable source of biofuels feedstock oil. Much of this research focuses on improving the overall per acre oil yield of Jatropha through advancements in genetics, soil science, and horticultural practices.

SG Biofuels, a San Diego-based jatropha developer, has used molecular breeding and biotechnology to produce elite hybrid seeds that show significant yield improvements over first-generation varieties. SG Biofuels also claims additional benefits have arisen from such strains, including improved flowering synchronicity, higher resistance to pests and diseases, and increased cold-weather tolerance.

Plant Research International, a department of the Wageningen University and Research Centre in the Netherlands, maintains an ongoing Jatropha Evaluation Project that examines the feasibility of large-scale jatropha cultivation through field and laboratory experiments. The Center for Sustainable Energy Farming (CfSEF) is a Los Angeles-based nonprofit research organization dedicated to jatropha research in the areas of plant science, agronomy, and horticulture. Successful exploration of these disciplines is projected to increase jatropha farm production yields by 200-300% in the next 10 years.

Fungi

A group at the Russian Academy of Sciences in Moscow, in a 2008 paper, stated they had isolated large amounts of lipids from single-celled fungi and turned it into biofuels in an economically efficient manner. More research on this fungal species, *Cunninghamella japonica*, and others, is likely to appear in the near future. The recent discovery of a variant of the fungus *Gliocladium roseum* (later renamed Ascocoryne sarcoides) points toward the production of so-called myco-diesel from cellulose. This organism was recently discovered in the rainforests of northern Patagonia, and has the unique capability of converting cellulose into medium-length hydrocarbons typically found in diesel fuel. Many other fungi that can degrade cellulose and other polymers have been observed to produce molecules that are currently being engineered using organisms from other kingdoms, suggesting that fungi may play a large role in the bio-production of fuels in the future (reviewed in).

Animal Gut Bacteria

Microbial gastrointestinal flora in a variety of animals have shown potential for the production of biofuels. Recent research has shown that TU-103, a strain of *Clostridium* bacteria found in Zebra feces, can convert nearly any form of cellulose into butanol fuel. Microbes in panda waste are being investigated for their use in creating biofuels from bamboo and other plant materials. There has also been substantial research into the technology of using the gut microbiomes of wood-feeding insects for the conversion of lignocellulotic material into biofuel.

Greenhouse Gas Emissions

Some scientists have expressed concerns about land-use change in response to greater demand for crops to use for biofuel and the subsequent carbon emissions. The payback period, that is, the time it will take biofuels to pay back the carbon debt they acquire due to land-use change, has been estimated to be between 100 and 1000 years, depending on the specific instance and location of land-use change. However, no-till practices combined with cover-crop practices can reduce the payback period to three years for grassland conversion and 14 years for forest conversion.

A study conducted in the Tocantis State, in northern Brazil, found that many families were cutting down forests in order to produce two conglomerates of oilseed plants, the J. curcas (JC group) and the R. communis (RC group). This region is composed of 15% Amazonian rainforest with high biodiversity, and 80% Cerrado forest with lower biodiversity. During the study, the farmers that planted the JC group released over 2193 Mg CO_2, while losing 53-105 Mg CO_2 sequestration from deforestation; and the RC group farmers released 562 Mg CO_2, while losing 48-90 Mg CO_2 to be sequestered from forest depletion. The production of these types of biofuels not only led into an increased emission of carbon dioxide, but also to lower efficiency of forests to absorb the gases that these farms were emitting. This has to do with the amount of fossil fuel the production of fuel crops involves. In addition, the intensive use of monocropping agriculture requires large amounts of water irrigation, as well as of fertilizers, herbicides and pesticides. This does not only lead to poisonous chemicals to disperse on water runoff, but also to the emission of nitrous oxide (NO_2) as a fertilizer byproduct, which is three hundred times more efficient in producing a greenhouse effect than carbon dioxide (CO_2).

Converting rainforests, peatlands, savannas, or grasslands to produce food crop–based biofuels in Brazil, Southeast Asia, and the United States creates a "biofuel carbon debt" by releasing 17 to 420 times more CO_2 than the annual greenhouse gas (GHG) reductions that these biofuels would provide by displacing fossil fuels. Biofuels made from waste biomass or from biomass grown on abandoned agricultural lands incur little to no carbon debt.

Water Use

In addition to water required to grow crops, biofuel facilities require significant process water.

Pellet Fuel

Pellet fuels (or pellets) are biofuels made from compressed organic matter or biomass. Pellets can be made from any one of five general categories of biomass: industrial waste and co-products, food waste, agricultural residues, energy crops, and virgin lumber. Wood pellets are the most common type of pellet fuel and are generally made from compacted sawdust and related industrial wastes from the milling of lumber, manufacture of wood products and furniture, and construction. Other industrial waste sources include empty fruit bunches, palm kernel shells, coconut shells, and tree tops and branches discarded during logging operations. So-called "black pellets" are made of biomass, refined to resemble hard coal and were developed to be used in existing coal-fired power plants. Pellets are categorized by their heating value, moisture and ash content, and dimensions. They can be used as fuels for power generation, commercial or residential heating, and cooking. Pellets are extremely dense and can be produced with a low moisture content (below 10%) that allows them to be burned with a very high combustion efficiency.

Wood pellets

Fuels for Heating
• Heating oil
• Wood pellet
• Kerosene
• Propane
• Natural gas
• Wood
• Coal

Further, their regular geometry and small size allow automatic feeding with very fine calibration. They can be fed to a burner by auger feeding or by pneumatic conveying. Their high density also permits compact storage and transport over long distance. They can be conveniently blown from a tanker to a storage bunker or silo on a customer's premises.

A broad range of pellet stoves, central heating furnaces, and other heating appliances have been developed and marketed since the mid-1980s. In 1997 fully automatic wood pellet boilers with similar comfort level as oil and gas boilers became available in Austria. With the surge in the price of fossil fuels since 2005, the demand for pellet heating has increased in Europe and North America, and a sizable industry is emerging. According to the International Energy Agency Task 40, wood pellet production has more than doubled between 2006 and 2010 to over 14 million tons. In a 2012 report, the Biomass Energy Resource Center says that it expects wood pellet production in North America to double again in the next five years.

Production

Pellets are produced by compressing the wood material which has first passed through a hammer mill to provide a uniform dough-like mass. This mass is fed to a press, where it is squeezed through a die having holes of the size required (normally 6 mm diameter, sometimes 8 mm or larger). The high pressure of the press causes the temperature of the wood to increase greatly, and the lignin plasticizes slightly, forming a natural "glue" that holds the pellet together as it cools.

Pellet truck being filled at a plant in Germany.

Pellets can be made from grass and other non-woody forms of biomass that do not contain lignin: distiller's dried grains (a brewing industry byproduct) can be added to provide the necessary durability. A 2005 news story from Cornell University News suggested that grass pellet production was more advanced in Europe than North America. It suggested the benefits of grass as a feedstock included its short growing time (70 days), and ease of cultivation and processing. The story quoted Jerry Cherney, an agriculture professor at the school, stating that grasses produce 96% of the heat of wood and that "any mixture of grasses can be used, cut in mid- to late summer, left in the field to leach out minerals, then baled and pelleted. Drying of the hay is not required for pelleting, making the cost of processing less than with wood pelleting." In 2012, the Department of Agriculture of Nova Scotia announced as a demonstration project conversion of an oil-fired boiler to grass pellets at a research facility.

Rice-husk fuel-pellets are made by compacting rice-husk obtained as by-product of rice-growing from the fields. It also has similar characteristics to the wood-pellets and more environment-friendly, as the raw material is a waste-product. The energy content is about 4-4.2 kcal/kg and moisture content is typically less than 10%. The size of pellets is generally kept to be about 6mm diameter and 25mm length in the form of a cylinder; though larger cylinder or briquette forms are not uncommon. It is much cheaper than similar energy-pellets and can be compacted/manufactured from the husk at the farm itself, using cheap machinery. They generally are more environment-friendly as compared to wood-pellets. In the regions of the world where wheat is the predominant food-crop, wheat husk can also be compacted to produce energy-pellets, with characteristics similar to rice-husk pellets.

A report by CORRIM (Consortium On Research on Renewable Industrial Material) for the Life-Cycle Inventory of Wood Pellet Manufacturing and Utilization estimates the energy required to dry, pelletize and transport pellets is less than 11% of the energy content of the pellets if using pre-dried industrial wood waste. If the pellets are made directly from forest material, it takes up to 18% of the energy to dry the wood and additional 8% for transportation and manufacturing energy. An environmental impact assessment of exported wood pellets by the Department of Chemical and Mineral Engineering, University of Bologna, Italy and the Clean Energy Research Centre, at the University of British Columbia, published in 2009, concluded that the energy consumed to ship Canadian wood pellets from Vancouver to Stockholm (15,500 km via the Panama Canal), is about 14% of the total energy content of the wood pellets.

Pellet Standards

Pellets conforming to the norms commonly used in Europe (DIN 51731 or Ö-Norm M-7135) have less than 10% water content, are uniform in density (higher than 1 ton per cubic meter, thus it sinks in water)(bulk density about 0.6-0.7 ton per cubic meter), have good structural strength, and low dust and ash content. Because the wood fibres are broken down by the hammer mill, there is virtually no difference in the finished pellets between different wood types. Pellets can be made from nearly any wood variety, provided the pellet press is equipped with good instrumentation, the differences in feed material can be compensated for in the press regulation.. In Europe, the main production areas are located in south Scandinavia, Finland, Central Europe, Austria, and the Baltic countries.

Pellets conforming to the European standards norms which contain recycled wood or outside contaminants are considered Class B pellets. Recycled materials such as particle board, treated or painted wood, melamine resin-coated panels and the like are particularly unsuitable for use in pellets, since they may produce noxious emissions and uncontrolled variations in the burning characteristics of the pellets.

Standards used in the United States are different, developed by the Pellet Fuels Institute and, as in Europe, are not mandatory. Still, many manufacturers comply, as warranties of US-manufactured or imported combustion equipment may not cover damage by pellets non-conformant with regulations. Prices for US pellets surged during the fossil fuel price inflation of 2007–2008, but later dropped markedly and are generally lower on a per-BTU basis than most fossil fuels, excluding coal.

Regulatory agencies in Europe and North America are in the process of tightening the emissions standards for all forms of wood heat, including wood pellets and pellet stoves. These standards will become mandatory, with independently certified testing to ensure compliance. In the United States, the new rules initiated in 2009 have completed the EPA regulatory review process, with final new Standards of Performance for New Residential Wood Heaters rules issued for comment on June 24, 2014. The American Lumber Standard Committee will be the independent certification agency for the new pellet standards.

Pellet Stove Operation

There are three general types of pellet heating appliances, free standing pellet stoves, pellet stove inserts and pellet boilers. *Pellet stoves* "look like traditional wood stoves but operate more like a modern furnace. [Fuel, wood or other biomass pellets, is stored in a storage bin called a hopper. The hopper can be located on the top of the appliance, the side of it or remotely.] A mechanical auger [automatically feeds] the pellets into a burn pot, where they are incinerated at such a high temperature that they create no vent-clogging creosote and very little ash or emissions... "Heat-exchange tubes": Send air heated by fire into room... "Convection fan": Circulates air through heat-exchange tubes and into room... The biggest difference between a pellet stove and ... a woodstove, is that, inside, the pellet stove is a high-tech device with a circuit board, a thermostat, and fans—all of which work together to [regulate temperature and] heat your space efficiently."

A *pellet stove insert* is a stove that is inserted into an existing masonry or prefabricated wood fireplace.

Pellet boilers are standalone central heating and hot water systems designed to replace traditional fossil fuel systems in residential, commercial and institutional applications.

Automatic or *auto-pellet boilers* include silos for bulk storage of pellets, a fuel delivery system that moves the fuel from the silo to the hopper, a logic controller to regulate temperature across multiple heating zones and an automated ash removal system for long-term automated operations.

Pellet baskets allow a person to heat their home using pellets in existing stoves or fire-places.

Energy Output and Efficiency

The energy content of wood pellets is approximately 4.7 – 5.2 MWh/tonne (~7450 BTU/lb).

Wood-pellet heater

High-efficiency wood pellet stoves and boilers have been developed in recent years, typically offering combustion efficiencies of over 85%. The newest generation of wood pellet boilers can work in condensing mode and therefore achieve 12% higher efficiency values. Wood pellet boilers have limited control over the rate and presence of combustion compared to liquid or gaseous-fired systems; however, for this reason they are better suited for hydronic heating systems due to the hydronic system's greater ability to store heat. Pellet burners capable of being retrofitted to oil-burning boilers are also available.

It was found that 1000 kilograms of wood pellets can give as much energy and heat as it can be received from 1600 kilograms of logwood, 475 cubic meters of gas, 500 liters of diesel fuel and 685 liters of fuel oil.

Air Pollution Emissions

Emissions such as NO_x, SO_x and volatile organic compounds from pellet burning equipment are in general very low in comparison to other forms of combustion heating. A recognized problem is the emission of fine particulate matter to the air, especially in urban areas that have a high concentration of pellet heating systems or coal or oil heating systems in close proximity. This $PM_{2.5}$ emissions of older pellet stoves and boilers can be problematic in close quarters, especially in comparison to natural gas (or renewable biogas), though on large installations electrostatic precipitators, cyclonic separators, or baghouse particle filters can control particulates when properly maintained and operated.

Global Warming

There is uncertainty to what degree making heat or electricity by burning wood pellets contributes to global climate change, as well as how the impact on climate compares to the impact of using competing sources of heat. Factors in the uncertainty include the wood source, carbon dioxide emissions from production and transport as well as from final combustion, and what time scale is appropriate for the consideration.

A report by the Manomet Center for Conservation Sciences, "Biomass Sustainability and Carbon Policy Study" issued in June 2010 for the Massachusetts Department of Energy Resources, concludes that burning biomass such as wood pellets or wood chips releases a large amount of CO_2 into the air, creating a "carbon debt" that is not retired for 20–25 years and after which there is a net benefit. In June 2011 the department was preparing to file its final regulation, expecting to significantly tighten controls on the use of biomass for energy, including wood pellets. Biomass energy proponents have disputed the Manomet report's conclusions, and scientists have pointed out oversights in the report, suggesting that climate impacts are worse than reported.

Until ca. 2008 it was commonly assumed, even in scientific papers, that biomass energy (including from wood pellets) is carbon neutral, largely because regrowth of vegetation was believed to recapture and store the carbon that is emitted to the air. Then, scientific papers studying the climate implications of biomass began to appear which refuted the simplistic assumption of its carbon neutrality. According to the Biomass Energy Resource Center, the assumption of carbon neutrality "has shifted to a recognition that the carbon implications of biomass depend on how the fuel is harvested, from what forest types, what kinds of forest management are applied, and how biomass is used over time and across the landscape."

In 2011 twelve prominent U.S. environmental organizations adopted policy setting a high bar for government incentives of biomass energy, including wood pellets. It states in part that, "[b]iomass sources and facilities qualifying for (government) incentives must result in lower life-cycle, cumulative and net GHG and ocean acidifying emissions, within 20 years and also over the longer term, than the energy sources they replace or compete with."

Sustainability

The wood products industry is concerned that if large-scale use of wood energy is instituted, the supply of raw materials for construction and manufacturing will be significantly curtailed.

Cost

Due to the rapid increase in popularity since 2005, pellet availability and cost may be an issue. This is an important consideration when buying a pellet stove, furnace, pellet baskets or other devices known in the industry as Bradley Burners. However, current pellet production is increasing and there are plans to bring several new pellet mills on-line in the US in 2008–2009.

The cost of the pellets can be affected by the building cycle leading to fluctuations in the supply of sawdust and offcuts.

Per the New Hampshire Office of Energy and Planning release on Fuel Prices updated on 5 Oct 2015, the cost of #2 Fuel Oil delivered can be compared to the cost of Bulk Delivered Wood Fuel Pellets using their BTU equivalent: 1 ton pellets = 118.97 gallon of #2 Fuel Oil. This assumes that one ton of pellets produces 16,500,000 BTU and one gallon of #2 Fuel Oil produces 138,690 BTU. Thus if #2 Fuel Oil delivered costs $1.90/Gal, the breakeven price for pellets is $238.00/Ton delivered.

Usage by Region

Europe

Usage across Europe varies due to government regulations. In the Netherlands, Belgium, and the UK, pellets are used mainly in large-scale power plants. In Denmark and Sweden, pellets are used in large-scale power plants, medium-scale district heating systems, and small-scale residential heat. In Germany, Austria, Italy, and France, pellets are used mostly for small-scale residential and industrial heat.

EU Pellet Use (ton)	
Country	**2013**
UK	4 540 000
Italy	3 300 000
Denmark	2 500 000
Netherlands	2 000 000
Sweden	1 650 000
Germany	1 600 000
Belgium	1 320 000

The UK has initiated a grant scheme called the Renewable Heat Incentive (RHI) allowing non-domestic and domestic wood pellet boiler installations to receive payments over a period of between 7–20 years It is the first such scheme in the world and aims to increase the amount of renewable energy generated in the UK, in line with EU commitments. Scotland and Northern Ireland have separate but similar schemes. From Spring 2015, any biomass owners whether domestic or commercial must buy their fuels from BSL (Biomass Suppliers List) approved suppliers in order to receive RHI payments.

Pellets are widely used in Sweden, the main pellet producer in Europe, mainly as an alternative to oil-fired central heating. In Austria, the leading market for pellet central heating furnaces (relative to its population), it is estimated that 2/3 of all new domestic heating furnaces are pellet burners. In Italy, a large market for automatically fed pellet stoves has developed. Italy's main usage for pellets is small - scale private residential and industrial boilers for heating.

In 2014 in Germany the overall wood pellet consumption per year comprised 2,2 mln tones. These pellets are consumed predominantly by residential small scale heating sector. The co-firing plants which use pellet sector for energy production are not wide-spread in the country. The largest amount of wood pellets is certified with DINplus and these are the pellets of the highest quality. As a rule, the pellets of lower quality are exported.

New Zealand

The total sales of wood pellets in New Zealand was 3–5,000 tonnes in 2003. Recent construction of new wood pellet plants has given a huge increase in production capacity.

United States

Some companies import European-made boilers. As of 2009, about 800,000 Americans were using wood pellets for heat. It is estimated that 2.33 million tons of wood pellets will be used for heat in the US in 2013. The US wood pellet export to Europe grew from 1.24 million ton in 2006 to 7 million ton in 2012, but forests grew even more.

Other Uses

Horse Bedding

When small amounts of water are added to wood pellets, they expand and revert to sawdust. This makes them suitable to use as a horse bedding. The ease of storage and transportation are additional benefits over traditional bedding. However, some species of wood, including walnut, can be toxic to horses and should never be used for bedding.

In Thailand, rice husk pellets are being produced for animal bedding. They have a high absorption rate which makes them ideal for the purpose.

Absorbents

Wood pellets are also used to absorb contaminated water when drilling oil or gas wells.

References

- "Publications - International Resource Panel". Archived from the original on 1 January 2016. Retrieved 30 May 2015.

- Overview of Greenhouse Gases, Methane Emissions. Climate Change, United States Environmental Protection Agency, 11 December 2015.

- Administrator. "Biogas CHP - Alfagy - Profitable Greener Energy via CHP, Cogen and Biomass Boiler using Wood, Biogas, Natural Gas, Biodiesel, Vegetable Oil, Syngas and Straw". Retrieved 15 May 2015.

- "Biogas plants provide cooking and fertiliser". Ashden Awards, sustainable and renewable energy in the UK and developing world. Retrieved 15 May 2015.

- Editor, Energy News (June 30, 2014). "Forest Fuels Secures Biomass Suppliers List Registration". OilFired Up. RuralEnergyNews.co.uk. Retrieved 2014-12-28.

- Summers, Rebecca (24 April 2013) Bacteria churn out first ever petrol-like biofuel New Scientist, Retrieved 27 April 2013.

- "Panda Poop Might Help Turn Plants Into Fuel". News.nationalgeographic.com. 2013-09-10. Retrieved 2013-10-02.

- Retka Schill, Sue (July 31, 2013). "PFI conference: Continued growth expected in pellet market". Biomass Magazine. Retrieved 16 September 2013.

- Ethanol Research (2012-04-02). "National Corn-to-Ethanol Research Center (NCERC)". Ethanol Research. Archived from the original on 20 March 2012. Retrieved 2012-04-02.

- American Coalition for Ethanol (2008-06-02). "Responses to Questions from Senator Bingaman" (PDF). American Coalition for Ethanol. Archived from the original (PDF) on 4 October 2011. Retrieved 2012-04-02.

- National Renewable Energy Laboratory (2007-03-02). "Research Advantages: Cellulosic Ethanol" (PDF). National Renewable Energy Laboratory. Retrieved 2012-04-02.

- Biofuels Digest (2011-05-16). "Jatropha blooms again: SG Biofuels secures 250K acres for hybrids". Biofuels Digest. Retrieved 2012-03-08.

- Plant Research International (2012-03-08). "JATROPT (Jatropha curcas): Applied and technical research into plant properties". Plant Research International. Retrieved 2012-03-08.

- Biofuels Magazine (2011-04-11). "Energy Farming Methods Mature, Improve". Biofuels Magazine. Retrieved 2012-03-08.

- Kathryn Hobgood Ray (August 25, 2011). "Cars Could Run on Recycled Newspaper, Tulane Scientists Say". Tulane University news webpage. Tulane University. Retrieved March 14, 2012.

- Cocchi, Maurizio (December 2011). "Global Wood Pellet Industry Market and Trade Study" (PDF). IEA Task 40. Retrieved 1 June 2012.

- Frederick, Paul. "2012 VT Wood Chip & Pellet Heating Conference" (PDF). Biomass Energy Resource Center. Retrieved 23 January 2012.

- "Revision of New Source Performance Standards for New Residential Wood Heaters". US EPA. 2011-10-26. Retrieved 2 January 2012.

- Donegan, Fran J. "Get More Heat from Your Fireplace and Wood Stove". This Old House. Retrieved 29 October 2012.

Biodiesel and its Environmental Impacts

Biodiesel usually refers to a vegetable oil; it is typically made by chemically reacting lipids. Biodiesel is used with the intention of reducing greenhouse gas. Whether it certainly does cause less environmental issues or not depends on many factors. The diverse applications of biodiesel in the current scenario have been thoroughly discussed in this section.

Biodiesel

Biodiesel refers to a vegetable oil - or animal fat-based diesel fuel consisting of long-chain alkyl (methyl, ethyl, or propyl) esters. Biodiesel is typically made by chemically reacting lipids (e.g., vegetable oil, soybean oil, animal fat (tallow)) with an alcohol producing fatty acid esters.

Biodiesel is meant to be used in standard diesel engines and is thus distinct from the vegetable and waste oils used to fuel *converted* diesel engines. Biodiesel can be used alone, or blended with petrodiesel in any proportions. Biodiesel blends can also be used as heating oil.

The National Biodiesel Board (USA) also has a technical definition of "biodiesel" as a mono-alkyl ester.

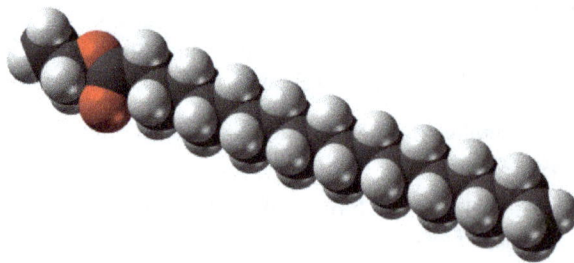

Space-filling model of ethyl stearate, or stearic acid ethyl ester, an ethyl ester produced from soybean or canola oil and ethanol

Space-filling model of methyl linoleate, or linoleic acid methyl ester,
a common methyl ester produced from soybean or canola oil and methanol

Blends

Blends of biodiesel and conventional hydrocarbon-based diesel are products most commonly distributed for use in the retail diesel fuel marketplace. Much of the world uses a system known as the "B" factor to state the amount of biodiesel in any fuel mix:

Biodiesel sample

- 100% biodiesel is referred to as B100

- 20% biodiesel, 80% petrodiesel is labeled B20

- 5% biodiesel, 95% petrodiesel is labeled B5

- 2% biodiesel, 98% petrodiesel is labeled B2

Blends of 20% biodiesel and lower can be used in diesel equipment with no, or only minor modifications, although certain manufacturers do not extend warranty coverage if equipment is damaged by these blends. The B6 to B20 blends are covered by the ASTM D7467 specification. Biodiesel can also be used in its pure form (B100), but may require certain engine modifications to avoid maintenance and performance problems. Blending B100 with petroleum diesel may be accomplished by:

- Mixing in tanks at manufacturing point prior to delivery to tanker truck

- Splash mixing in the tanker truck (adding specific percentages of biodiesel and petroleum diesel)

- In-line mixing, two components arrive at tanker truck simultaneously.

- Metered pump mixing, petroleum diesel and biodiesel meters are set to X total volume, transfer pump pulls from two points and mix is complete on leaving pump.

Applications

Biodiesel can be used in pure form (B100) or may be blended with petroleum diesel at any concentration in most injection pump diesel engines. New extreme high-pressure (29,000 psi) common rail engines have strict factory limits of B5 or B20, depending on manufacturer. Biodiesel has different solvent properties than petrodiesel, and will degrade natural rubber gaskets and hoses in vehicles (mostly vehicles manufactured before 1992), although these tend to wear out naturally and most likely will have already been replaced with FKM, which is nonreactive to biodiesel. Biodiesel has been known to break down deposits of residue in the fuel lines where petrodiesel has been used. As a result, fuel filters may become clogged with particulates if a quick transition to pure biodiesel is made. Therefore, it is recommended to change the fuel filters on engines and heaters shortly after first switching to a biodiesel blend.

Distribution

Since the passage of the Energy Policy Act of 2005, biodiesel use has been increasing in the United States. In the UK, the Renewable Transport Fuel Obligation obliges suppliers to include 5% renewable fuel in all transport fuel sold in the UK by 2010. For road diesel, this effectively means 5% biodiesel (B5).

Vehicular Use and Manufacturer Acceptance

In 2005, Chrysler (then part of DaimlerChrysler) released the Jeep Liberty CRD diesels from the factory into the European market with 5% biodiesel blends, indicating at least partial acceptance of biodiesel as an acceptable diesel fuel additive. In 2007, Daimler-Chrysler indicated its intention to increase warranty coverage to 20% biodiesel blends if biofuel quality in the United States can be standardized.

The Volkswagen Group has released a statement indicating that several of its vehicles are compatible with B5 and B100 made from rape seed oil and compatible with the EN 14214 standard. The use of the specified biodiesel type in its cars will not void any warranty.

Mercedes Benz does not allow diesel fuels containing greater than 5% biodiesel (B5) due to concerns about "production shortcomings". Any damages caused by the use of such non-approved fuels will not be covered by the Mercedes-Benz Limited Warranty.

Starting in 2004, the city of Halifax, Nova Scotia decided to update its bus system to allow the fleet of city buses to run entirely on a fish-oil based biodiesel. This caused the city some initial mechanical issues, but after several years of refining, the entire fleet had successfully been converted.

In 2007, McDonalds of UK announced it would start producing biodiesel from the waste oil byproduct of its restaurants. This fuel would be used to run its fleet.

The 2014 Chevy Cruze Clean Turbo Diesel, direct from the factory, will be rated for up to B20 (blend of 20% biodiesel / 80% regular diesel) biodiesel compatibility

Railway Usage

British train operating company Virgin Trains claimed to have run the UK's first "bio-diesel train", which was converted to run on 80% petrodiesel and 20% biodiesel.

Biodiesel locomotive and its external fuel tank at Mount Washington Cog Railway

The Royal Train on 15 September 2007 completed its first ever journey run on 100% biodiesel fuel supplied by Green Fuels Ltd. His Royal Highness, The Prince of Wales, and Green Fuels managing director, James Hygate, were the first passengers on a train fueled entirely by biodiesel fuel. Since 2007, the Royal Train has operated successfully on B100 (100% biodiesel).

Similarly, a state-owned short-line railroad in eastern Washington ran a test of a 25% bio-diesel / 75% petrodiesel blend during the summer of 2008, purchasing fuel from a biodiesel producer sited along the railroad tracks. The train will be powered by biodiesel made in part from canola grown in agricultural regions through which the short line runs.

Also in 2007, Disneyland began running the park trains on B98 (98% biodiesel). The program was discontinued in 2008 due to storage issues, but in January 2009, it was announced that the park would then be running all trains on biodiesel manufactured from its own used cooking oils. This is a change from running the trains on soy-based biodiesel.

In 2007, the historic Mt. Washington Cog Railway added the first biodiesel locomotive to its all-steam locomotive fleet. The fleet has climbed up the western slopes of Mount Washington in New Hampshire since 1868 with a peak vertical climb of 37.4 degrees.

On 8 July 2014, the then Indian Railway Minister D.V. Sadananda Gowda announced in Railway Budget that 5% bio-diesel will be used in Indian Railways' Diesel Engines.

Aircraft Use

A test flight has been performed by a Czech jet aircraft completely powered on bio-diesel. Other recent jet flights using biofuel, however, have been using other types of renewable fuels.

On November 7, 2011 United Airlines flew the world's first commercial aviation flight on a microbially derived biofuel using Solajet™, Solazyme's algae-derived renewable jet fuel. The Eco-skies Boeing 737-800 plane was fueled with 40 percent Solajet and 60 percent petroleum-derived jet fuel. The commercial Eco-skies flight 1403 departed from Houston's IAH airport at 10:30 and landed at Chicago's ORD airport at 13:03.

In September 2016, the Dutch flag carrier KLM contracted AltAir Fuels to supply all KLM flights departing Los Angeles International Airport with biofuel. For the next three years the Paramount, California based company will pump biofuel directly to the airport from their nearby refinery.

As a Heating Oil

Biodiesel can also be used as a heating fuel in domestic and commercial boilers, a mix of heating oil and biofuel which is standardized and taxed slightly differently from diesel fuel used for transportation. Bioheat® fuel is a proprietary blend of biodiesel and traditional heating oil. Bioheat® is a registered trademark of the National Biodiesel Board [NBB] and the National Oilheat Research Alliance [NORA] in the U.S., and Columbia Fuels in Canada). Heating biodiesel is available in various blends. ASTM 396 recognizes blends of up to 5 percent biodiesel as equivalent to pure petroleum heating oil. Blends of higher levels of up to 20% biofuel are used by many consumers. Research is underway to determine whether such blends affect performance.

Older furnaces may contain rubber parts that would be affected by biodiesel's solvent properties, but can otherwise burn biodiesel without any conversion required. Care must be taken, however, given that varnishes left behind by petrodiesel will be released and can clog pipes- fuel filtering and prompt filter replacement is required. Another approach is to start using biodiesel as a blend, and decreasing the petroleum proportion over time can allow the varnishes to come off more gradually and be less likely to clog. Thanks to its strong solvent properties, however, the furnace is cleaned out and generally becomes more efficient. A technical research paper describes laboratory research and field trials project using pure biodiesel and biodiesel blends as a heating fuel in oil-fired boilers. During the Biodiesel Expo 2006 in the UK, Andrew J. Robertson presented his biodiesel heating oil research from his technical paper and suggested B20 biodiesel could reduce UK household CO_2 emissions by 1.5 million tons per year.

A law passed under Massachusetts Governor Deval Patrick requires all home heating diesel in that state to be 2% biofuel by July 1, 2010, and 5% biofuel by 2013. New York City has passed a similar law.

Cleaning Oil Spills

With 80-90% of oil spill costs invested in shoreline cleanup, there is a search for more efficient and cost-effective methods to extract oil spills from the shorelines. Biodiesel has displayed its capacity to significantly dissolve crude oil, depending on the source of the fatty acids. In a laboratory setting, oiled sediments that simulated polluted shorelines were sprayed with a single coat of biodiesel and exposed to simulated tides. Biodiesel is an effective solvent to oil due to its methyl ester component, which considerably lowers the viscosity of the crude oil. Additionally, it has a higher buoyancy than crude oil, which later aids in its removal. As a result, 80% of oil was removed from cobble and fine sand, 50% in coarse sand, and 30% in gravel. Once the oil is liberated from the shoreline, the oil-biodiesel mixture is manually removed from the water surface with skimmers. Any remaining mixture is easily broken down due to the high biodegradability of biodiesel, and the increased surface area exposure of the mixture.

Biodiesel in Generators

Biodiesel is also used in rental generators

In 2001, UC Riverside installed a 6-megawatt backup power system that is entirely fueled by biodiesel. Backup diesel-fueled generators allow companies to avoid damaging blackouts of critical operations at the expense of high pollution and emission rates. By using B100, these generators were able to essentially eliminate the byproducts that result in smog, ozone, and sulfur emissions. The use of these generators in residential areas around schools, hospitals, and the general public result in substantial reductions in poisonous carbon monoxide and particulate matter.

Historical Background

Transesterification of a vegetable oil was conducted as early as 1853 by Patrick Duffy, four decades before the first diesel engine became functional. Rudolf Diesel's prime model, a single 10 ft (3.0 m) iron cylinder with a flywheel at its base, ran on its own power for the first time in Augsburg, Germany, on 10 August 1893 running on nothing but peanut oil. In remembrance of this event, 10 August has been declared "International Biodiesel Day".

Rudolf Diesel

It is often reported that Diesel designed his engine to run on peanut oil, but this is not the case. Diesel stated in his published papers, "at the Paris Exhibition in 1900 (*Exposition Universelle*) there was shown by the Otto Company a small Diesel engine, which, at the request of the French government ran on arachide (earth-nut or peanut) oil, and worked so smoothly that only a few people were aware of it. The engine was constructed for using mineral oil, and was then worked on vegetable oil without any alterations being made. The French Government at the time thought of testing the applicability to power production of the Arachide, or earth-nut, which grows in considerable quantities in their African colonies, and can easily be cultivated there." Diesel himself later conducted related tests and appeared supportive of the idea. In a 1912 speech Diesel said, "the use of vegetable oils for engine fuels may seem insignificant today but such oils may become, in the course of time, as important as petroleum and the coal-tar products of the present time."

Despite the widespread use of petroleum-derived diesel fuels, interest in vegetable oils as fuels for internal combustion engines was reported in several countries during the 1920s and 1930s and later during World War II. Belgium, France, Italy, the United Kingdom, Portugal, Germany, Brazil, Argentina, Japan and China were reported to have tested and used vegetable oils as diesel fuels during this time. Some operational problems were reported due to the high viscosity of vegetable oils compared to petroleum diesel fuel, which results in poor atomization of the fuel in the fuel spray and often leads to deposits and coking of the injectors, combustion chamber and valves. Attempts to overcome these problems included heating of the vegetable oil, blending it with petroleum-derived diesel fuel or ethanol, pyrolysis and cracking of the oils.

On 31 August 1937, G. Chavanne of the University of Brussels (Belgium) was granted a patent for a "Procedure for the transformation of vegetable oils for their uses as fuels" (fr. "*Procédé de Transformation d'Huiles Végétales en Vue de Leur Utilisation comme Carburants*") Belgian Patent 422,877. This patent described the alcoholysis (often referred to as transesterification) of vegetable oils using ethanol (and mentions methanol) in order to separate the fatty acids from the glycerol by replacing the glycerol with short linear alcohols. This appears to be the first account of the production of what is known as "biodiesel" today.

More recently, in 1977, Brazilian scientist Expedito Parente invented and submitted for patent, the first industrial process for the production of biodiesel. This process is classified as biodiesel by international norms, conferring a "standardized identity and quality. No other proposed biofuel has been validated by the motor industry." As of 2010, Parente's company Tecbio is working with Boeing and NASA to certify bioquerosene (bio-kerosene), another product produced and patented by the Brazilian scientist.

Research into the use of transesterified sunflower oil, and refining it to diesel fuel standards, was initiated in South Africa in 1979. By 1983, the process for producing fuel-quality, engine-tested biodiesel was completed and published internationally. An Austrian company, Gaskoks, obtained the technology from the South African Agricultural Engineers; the company erected the first biodiesel pilot plant in November 1987, and the first industrial-scale plant in April 1989 (with a capacity of 30,000 tons of rapeseed per annum).

Throughout the 1990s, plants were opened in many European countries, including the Czech Republic, Germany and Sweden. France launched local production of biodiesel fuel (referred to as *diester*) from rapeseed oil, which is mixed into regular diesel fuel at a level of 5%, and into the diesel fuel used by some captive fleets (e.g. public transportation) at a level of 30%. Renault, Peugeot and other manufacturers have certified truck engines for use with up to that level of partial biodiesel; experiments with 50% biodiesel are underway. During the same period, nations in other parts of the world also saw local production of biodiesel starting up: by 1998, the Austrian Biofuels Institute had identified 21 countries with commercial biodiesel projects. 100% biodiesel is now available at many normal service stations across Europe.

Properties

Biodiesel has promising lubricating properties and cetane ratings compared to low sulfur diesel fuels. Depending on the engine, this might include high pressure injection pumps, pump injectors (also called *unit injectors*) and fuel injectors.

Older diesel Mercedes are popular for running on biodiesel.

The calorific value of biodiesel is about 37.27 MJ/kg. This is 9% lower than regular Number 2 petrodiesel. Variations in biodiesel energy density is more dependent on the feedstock used than the production process. Still, these variations are less than for petrodiesel. It has been claimed biodiesel gives better lubricity and more complete combustion thus increasing the engine energy output and partially compensating for the higher energy density of petrodiesel.

The color of biodiesel ranges from golden to dark brown, depending on the production method. It is slightly miscible with water, has a high boiling point and low vapor pressure. *The flash point of biodiesel (>130 °C, >266 °F) is significantly higher than that of petroleum diesel (64 °C, 147 °F) or gasoline (−45 °C, -52 °F). Biodiesel has a density of ~ 0.88 g/cm³, higher than petrodiesel (~ 0.85 g/cm³).

Biodiesel contains virtually no sulfur, and it is often used as an additive to Ultra-Low Sulfur Diesel (ULSD) fuel to aid with lubrication, as the sulfur compounds in petrodiesel provide much of the lubricity.

Fuel Efficiency

The power output of biodiesel depends on its blend, quality, and load conditions under which the fuel is burnt. The thermal efficiency for example of B100 as compared to B20 will vary due to the differing energy content of the various blends. Thermal efficiency of a fuel is based in part on fuel characteristics such as: viscosity, specific density, and flash point; these characteristics will change as the blends as well as the quality of biodiesel varies. The American Society for Testing and Materials has set standards in order to judge the quality of a given fuel sample.

Regarding brake thermal efficiency one study found that B40 was superior to traditional counterpart at higher compression ratios (this higher brake thermal efficiency was recorded at compression ratios of 21:1). It was noted that, as the compression ratios increased, the efficiency of all fuel types - as well as blends being tested - increased; though it was found that a blend of B40 was the most economical at a compression ratio of 21:1 over all other blends. The study implied that this increase in efficiency was due to fuel density, viscosity, and heating values of the fuels.

Combustion

Fuel systems on the modern diesel engine were not designed to accommodate biodiesel, while many heavy duty engines are able to run with biodiesel blends e.g. B20. Traditional direct injection fuel systems operate at roughly 3,000 psi at the injector tip while the modern common rail fuel system operates upwards of 30,000 PSI at the injector tip. Components are designed to operate at a great temperature range, from below freezing to over 1,000 degrees Fahrenheit. Diesel fuel is expected to burn efficiently and produce as few emissions as possible. As emission standards are being introduced to diesel engines the need to control harmful emissions is being designed into the parameters of diesel engine fuel systems. The traditional inline injection system is more forgiving to poorer quality fuels as opposed to the common rail fuel system. The higher pressures and tighter tolerances of the common rail system allows for greater control over atomization and injection timing. This control of atomization as well as combustion allows for greater efficiency of modern diesel engines as well as greater control over emissions. Components within a diesel fuel system interact with the fuel in a way to ensure efficient operation of the fuel system and so the engine. If an out-of-specification fuel is introduced to a system that has specific parameters of operation, then the integrity of the overall fuel system may be compromised. Some of these parameters such as spray pattern and atomization are directly related to injection timing.

One study found that during atomization biodiesel and its blends produced droplets were greater in diameter than the droplets produced by traditional petrodiesel. The smaller droplets were attributed to the lower viscosity and surface tension of traditional petrol. It was found that droplets at the periphery of the spray pattern were larger in diameter than the droplets at the center this was attributed to the faster pressure drop at the edge of the spray pattern; there was a proportional relationship between the droplet size and the distance from the injector tip. It was found that B100 had the greatest spray penetration, this was attributed to the greater density of B100. Having a greater droplet size can lead to; inefficiencies in the combustion, increased emissions, and decreased horse power. In another study it was found that there is a short injection delay when injecting biodiesel. This injection delay was attributed to the greater viscosity of Biodiesel. It was noted that the higher viscosity and the greater cetane rating of biodiesel over traditional petrodiesel lead to poor atomization, as well as mixture penetration with air during the ignition delay period. Another study noted that this ignition delay may aid in a decrease of NOx emission.

Emissions

Emissions are inherent to the combustion of diesel fuels that are regulated by the U.S. Environmental Protection Agency (E.P.A.). As these emissions are a byproduct of the combustion process, in order to ensure E.P.A. compliance a fuel system must be capable of controlling the combustion of fuels as well as the mitigation of emissions. There are a number of new technologies being phased in to control the production of diesel emissions. The exhaust gas recirculation system, E.G.R., and the diesel particulate filter, D.P.F., are both designed to mitigate the production of harmful emissions.

A study performed by the Chonbuk National University concluded that a B30 biodiesel blend reduced carbon monoxide emissions by approximately 83% and particulate matter emissions by roughly 33%. NOx emissions, however, were found to increase without the application of an E.G.R. system. The study also concluded that, with E.G.R, a B20 biodiesel blend considerably reduced the emissions of the engine. Additionally, analysis by the California Air Resources Board found that biodiesel had the lowest carbon emissions of the fuels tested, those being ultra-low-sulfur diesel, gasoline, corn-based ethanol, compressed natural gas, and five types of biodiesel from varying feedstocks. Their conclusions also showed great variance in carbon emissions of biodiesel based on the feedstock used. Of soy, tallow, canola, corn, and used cooking oil, soy showed the highest carbon emissions, while used cooking oil produced the lowest.

While studying the effect of biodiesel on a D.P.F. it was found that though the presence of sodium and potassium carbonates aided in the catalytic conversion of ash, as the diesel particulates are catalyzed, they may congregate inside the D.P.F. and so interfere with the clearances of the filter. This may cause the filter to clog and interfere with the regeneration process. In a study on the impact of E.G.R. rates with blends of jathropa biodiesel it was shown that there was a decrease in fuel efficiency and torque output due to the use of biodiesel on a diesel engine designed with an E.G.R. system. It was found that CO and CO_2 emissions increased with an increase in exhaust gas recirculation but NOx levels decreased. The opacity level of the jathropa blends was in an acceptable range, where traditional diesel was out of acceptable standards. It was shown that a decrease in Nox emissions could be obtained with an E.G.R. system. This study showed an advantage over traditional diesel within a certain operating range of the E.G.R. system. Currently blended biodiesel fuels (B5 and B20) are being used in many heavy-duty vehicles especially transit buses in US cities. Characterization of exhaust emissions showed significant emission reductions compared to regular diesel.

Material Compatibility

- Plastics: High-density polyethylene (HDPE) is compatible but polyvinyl chloride (PVC) is slowly degraded. Polystyrene is dissolved on contact with biodiesel.

- Metals: Biodiesel (like methanol) has an effect on copper-based materials (e.g. brass), and it also affects zinc, tin, lead, and cast iron. Stainless steels (316 and 304) and aluminum are unaffected.

- Rubber: Biodiesel also affects types of natural rubbers found in some older engine components. Studies have also found that fluorinated elastomers (FKM) cured with peroxide and base-metal oxides can be degraded when biodiesel loses its stability caused by oxidation. Commonly used synthetic rubbers FKM- GBL-S and FKM-GF-S found in modern vehicles were found to handle biodiesel in all conditions.

Technical Standards

Biodiesel has a number of standards for its quality including European standard EN 14214, ASTM International D6751, and others.

Low Temperature Gelling

When biodiesel is cooled below a certain point, some of the molecules aggregate and form crystals. The fuel starts to appear cloudy once the crystals become larger than one quarter of the wavelengths of visible light - this is the cloud point (CP). As the fuel is cooled further these crystals become larger. The lowest temperature at which fuel can pass through a 45 micrometre filter is the cold filter plugging point (CFPP). As biodiesel is cooled further it will gel and then solidify. Within Europe, there are differences in the CFPP requirements between countries. This is reflected in the different national standards of those countries. The temperature at which pure (B100) biodiesel starts to gel varies significantly and depends upon the mix of esters and therefore the feedstock oil used to produce the biodiesel. For example, biodiesel produced from low erucic acid varieties of canola seed (RME) starts to gel at approximately −10 °C (14 °F). Biodiesel produced from beef tallow and palm oil tends to gel at around 16 °C (61 °F) and 13 °C (55 °F) respectively. There are a number of commercially available additives that will significantly lower the pour point and cold filter plugging point of pure biodiesel. Winter operation is also possible by blending biodiesel with other fuel oils including #2 low sulfur diesel fuel and #1 diesel / kerosene.

Another approach to facilitate the use of biodiesel in cold conditions is by employing a second fuel tank for biodiesel in addition to the standard diesel fuel tank. The second fuel tank can be insulated and a heating coil using engine coolant is run through the tank. The fuel tanks can be switched over when the fuel is sufficiently warm. A similar method can be used to operate diesel vehicles using straight vegetable oil.

Contamination by Water

Biodiesel may contain small but problematic quantities of water. Although it is only slightly miscible with water it is hygroscopic. One of the reasons biodiesel can absorb water is the persistence of mono and diglycerides left over from an incomplete reaction. These molecules can act as an emulsifier, allowing water to mix with the biodiesel. In addition, there may be water that is residual to processing or resulting from storage tank condensation. The presence of water is a problem because:

- Water reduces the heat of fuel combustion, causing smoke, harder starting, and reduced power.

- Water causes corrosion of fuel system components (pumps, fuel lines, etc.)

- Microbes in water cause the paper-element filters in the system to rot and fail, causing failure of the fuel pump due to ingestion of large particles.

- Water freezes to form ice crystals that provide sites for nucleation, accelerating gelling of the fuel.

- Water causes pitting in pistons.

Previously, the amount of water contaminating biodiesel has been difficult to measure by taking samples, since water and oil separate. However, it is now possible to measure the water content using water-in-oil sensors.

Water contamination is also a potential problem when using certain chemical catalysts involved in the production process, substantially reducing catalytic efficiency of base (high pH) catalysts such as potassium hydroxide. However, the super-critical methanol production methodology, whereby the transesterification process of oil feedstock and methanol is effectuated under high temperature and pressure, has been shown to be largely unaffected by the presence of water contamination during the production phase.

Availability and Prices

Global biodiesel production reached 3.8 million tons in 2005. Approximately 85% of biodiesel production came from the European Union.

In some countries biodiesel is less expensive than conventional diesel

In 2007, in the United States, average retail (at the pump) prices, including federal and state fuel taxes, of B2/B5 were lower than petroleum diesel by about 12 cents, and B20 blends were the same as petrodiesel. However, as part of a dramatic shift in diesel pricing, by July 2009, the US DOE was reporting average costs of B20 15 cents per gallon higher than petroleum diesel ($2.69/gal vs. $2.54/gal). B99 and B100 generally cost more than petrodiesel except where local governments provide a tax incentive or subsidy.

Production

Biodiesel is commonly produced by the transesterification of the vegetable oil or animal fat feedstock. There are several methods for carrying out this transesterification reaction including the common batch process, supercritical processes, ultrasonic methods, and even microwave methods.

Chemically, transesterified biodiesel comprises a mix of mono-alkyl esters of long chain fatty acids. The most common form uses methanol (converted to sodium methoxide) to produce methyl esters (commonly referred to as Fatty Acid Methyl Ester - FAME) as it is the cheapest alcohol available, though ethanol can be used to produce an ethyl ester (commonly referred to as Fatty Acid Ethyl Ester - FAEE) biodiesel and higher alcohols such as isopropanol and butanol have also been used. Using alcohols of higher molecular weights improves the cold flow properties of the resulting ester, at the cost of a less efficient transesterification reaction. A lipid transesterification production process is used to convert the base oil to the desired esters. Any free fatty acids (FFAs) in the base oil are either converted to soap and removed from the process, or they are esterified (yielding more biodiesel) using an acidic catalyst. After this processing, unlike straight vegetable oil, biodiesel has combustion properties very similar to those of petroleum diesel, and can replace it in most current uses.

The methanol used in most biodiesel production processes is made using fossil fuel inputs. However, there are sources of renewable methanol made using carbon dioxide or biomass as feedstock, making their production processes free of fossil fuels.

A by-product of the transesterification process is the production of glycerol. For every 1 tonne of biodiesel that is manufactured, 100 kg of glycerol are produced. Originally, there was a valuable market for the glycerol, which assisted the economics of the process as a whole. However, with the increase in global biodiesel production, the market price for this crude glycerol (containing 20% water and catalyst residues) has crashed. Research is being conducted globally to use this glycerol as a chemical building block. One initiative in the UK is The Glycerol Challenge.

Usually this crude glycerol has to be purified, typically by performing vacuum distillation. This is rather energy intensive. The refined glycerol (98%+ purity) can then be utilised directly, or converted into other products. The following announcements were made in 2007: A joint venture of Ashland Inc. and Cargill announced plans to make propylene glycol in Europe from glycerol and Dow Chemical announced similar plans for North America. Dow also plans to build a plant in China to make epichlorhydrin from glycerol. Epichlorhydrin is a raw material for epoxy resins.

Production Levels

In 2007, biodiesel production capacity was growing rapidly, with an average annual growth rate from 2002-06 of over 40%. For the year 2006, the latest for which actual production figures could be obtained, total world biodiesel production was about 5-6 million tonnes, with 4.9 million tonnes processed in Europe (of which 2.7 million tonnes was from Germany) and most of the rest from the USA. In 2008 production in Europe alone had risen to 7.8 million tonnes. In July 2009, a duty was added to American imported biodiesel in the European Union in order to balance the competition from European, especially German producers. The capacity for 2008 in Europe totalled 16 million tonnes. This compares with a total demand for diesel in the US and Europe of approximately 490 million tonnes (147 billion gallons). Total world production of vegetable oil for all purposes in 2005/06 was about 110 million tonnes, with about 34 million tonnes each of palm oil and soybean oil.

US biodiesel production in 2011 brought the industry to a new milestone. Under the EPA Renewable Fuel Standard, targets have been implemented for the biodiesel production plants in order to monitor and document production levels in comparison to total demand. According to the year-end data released by the EPA, biodiesel production in 2011 reached more than 1 billion gallons. This production number far exceeded the 800 million gallon target set by the EPA. The projected production for 2020 is nearly 12 billion gallons.

Biodiesel Feedstocks

A variety of oils can be used to produce biodiesel. These include:

- Virgin oil feedstock – rapeseed and soybean oils are most commonly used, soybean oil accounting for about half of U.S. production. It also can be obtained from Pongamia, field pennycress and jatropha and other crops such as mustard, jojoba, flax, sunflower, palm oil, coconut and hemp;

- Waste vegetable oil (WVO);

- Animal fats including tallow, lard, yellow grease, chicken fat, and the by-products of the production of Omega-3 fatty acids from fish oil.

- Algae, which can be grown using waste materials such as sewage and without displacing land currently used for food production.

- Oil from halophytes such as *Salicornia bigelovii*, which can be grown using saltwater in coastal areas where conventional crops cannot be grown, with yields equal to the yields of soybeans and other oilseeds grown using freshwater irrigation

- Sewage Sludge - The sewage-to-biofuel field is attracting interest from major companies like Waste Management and startups like InfoSpi, which are betting that renewable sewage biodiesel can become competitive with petroleum diesel on price.

Many advocates suggest that waste vegetable oil is the best source of oil to produce biodiesel, but since the available supply is drastically less than the amount of petroleum-based fuel that is burned for transportation and home heating in the world, this local solution could not scale to the current rate of consumption.

Animal fats are a by-product of meat production and cooking. Although it would not be efficient to raise animals (or catch fish) simply for their fat, use of the by-product adds value to the livestock industry (hogs, cattle, poultry). Today, multi-feedstock biodiesel facilities are producing high quality animal-fat based biodiesel. Currently, a 5-million dollar plant is being built in the USA, with the intent of producing 11.4 million litres (3 million gallons) biodiesel from some of the estimated 1 billion kg (2.2 billion pounds) of chicken fat produced annually at the local Tyson poultry plant. Similarly, some small-scale biodiesel factories use waste fish oil as feedstock. An EU-funded project (ENERFISH) suggests that at a Vietnamese plant to produce biodiesel from catfish (basa, also known as pangasius), an output of 13 tons/day of biodiesel can be produced from 81 tons of fish waste (in turn resulting from 130 tons of fish). This project utilises the biodiesel to fuel a CHP unit in the fish processing plant, mainly to power the fish freezing plant.

Quantity of Feedstocks Required

Current worldwide production of vegetable oil and animal fat is not sufficient to replace liquid fossil fuel use. Furthermore, some object to the vast amount of farming and the resulting fertilization, pesticide use, and land use conversion that would be needed to produce the additional vegetable oil. The estimated transportation diesel fuel and home heating oil used in the United States is about 160 million tons (350 billion pounds) according to the Energy Information Administration, US Department of Energy. In the United States, estimated production of vegetable oil for all uses is about 11 million tons (24 billion pounds) and estimated production of animal fat is 5.3 million tonnes (12 billion pounds).

If the entire arable land area of the USA (470 million acres, or 1.9 million square kilometers) were devoted to biodiesel production from soy, this would just about provide the 160 million tonnes required (assuming an optimistic 98 US gal/acre of biodiesel). This land area could in principle be reduced significantly using algae, if the obstacles can be overcome. The US DOE estimates that if algae fuel replaced all the petroleum fuel in the United States, it would require 15,000 square miles (39,000 square kilometers), which is a few thousand square miles larger than Maryland, or 30% greater than the area of Belgium, assuming a yield of 140 tonnes/hectare (15,000 US gal/acre). Given a more realistic yield of 36 tonnes/hectare (3834 US gal/acre) the area required is about 152,000 square kilometers, or roughly equal to that of the state of Georgia or of England and Wales. The advantages of algae are that it can be grown on non-arable land such as deserts or in marine environments, and the potential oil yields are much higher than from plants.

Yield

Feedstock yield efficiency per unit area affects the feasibility of ramping up production to the huge industrial levels required to power a significant percentage of vehicles.

Some typical yields		
Crop	**Yield**	
	L/ha	**US gal/acre**
Palm oil	4752	508
Coconut	2151	230
Cyperus esculentus	1628	174
Rapeseed	954	102
Soy (Indiana)	554-922	59.2-98.6
Chinese tallow	907	97
Peanut	842	90
Sunflower	767	82
Hemp	242	26

1. "Biofuels: some numbers". Grist.org. *Retrieved 2010-03-15.*

2. Makareviciene et al., "Opportunities for the use of chufa sedge in biodiesel production", Industrial Crops and Products, 50 (2013) p. 635, table 2.

3. Klass, Donald, "Biomass for Renewable Energy, Fuels, and Chemicals", page 341. Academic Press, 1998.

4. Kitani, Osamu, "Volume V: Energy and Biomass Engineering, CIGR Handbook of Agricultural Engineering", Amer Society of Agricultural, 1999.

Algae fuel yields have not yet been accurately determined, but DOE is reported as saying that algae yield 30 times more energy per acre than land crops such as soybeans. Yields of 36 tonnes/hectare are considered practical by Ami Ben-Amotz of the Institute of Oceanography in Haifa, who has been farming Algae commercially for over 20 years.

Jatropha has been cited as a high-yield source of biodiesel but yields are highly dependent on climatic and soil conditions. The estimates at the low end put the yield at about 200 US gal/acre (1.5-2 tonnes per hectare) per crop; in more favorable climates two or more crops per year have been achieved. It is grown in the Philippines, Mali and India, is drought-resistant, and can share space with other cash crops such as coffee, sugar, fruits and vegetables. It is well-suited to semi-arid lands and can contribute to slow down desertification, according to its advocates.

Efficiency and Economic Arguments

According to a study by Drs. Van Dyne and Raymer for the Tennessee Valley Authority, the average US farm consumes fuel at the rate of 82 litres per hectare (8.75 US gal/acre) of land to produce one crop. However, average crops of rapeseed produce oil at an average rate of 1,029 L/ha (110 US gal/acre), and high-yield rapeseed fields produce about 1,356 L/ha (145 US gal/acre). The ratio of input to output in these cases is roughly 1:12.5 and 1:16.5. Photosynthesis is known to have an efficiency rate of about 3-6% of total solar radiation and if the entire mass of a crop is utilized for energy production, the overall efficiency of this chain is currently about 1% While this may compare unfavorably to solar cells combined with an electric drive train, biodiesel is less costly to deploy (solar cells cost approximately US$250 per square meter) and transport (electric vehicles require batteries which currently have a much lower energy density than liquid fuels). A 2005 study found that biodiesel production using soybeans required 27% more fossil energy than the biodiesel produced and 118% more energy using sunflowers.

Pure biodiesel (B-100) made from soybeans

However, these statistics by themselves are not enough to show whether such a change makes economic sense. Additional factors must be taken into account, such as: the fuel equivalent of the energy required for processing, the yield of fuel from raw oil, the return on cultivating food, the effect biodiesel will have on food prices and the relative cost of biodiesel versus petrodiesel, water pollution from farm run-off, soil depletion, and the externalized costs of political and military interference in oil-producing countries intended to control the price of petrodiesel.

The debate over the energy balance of biodiesel is ongoing. Transitioning fully to biofuels could require immense tracts of land if traditional food crops are used (although non food crops can be utilized). The problem would be especially severe for nations with large economies, since energy consumption scales with economic output.

If using only traditional food plants, most such nations do not have sufficient arable land to produce biofuel for the nation's vehicles. Nations with smaller economies (hence less energy consumption) and more arable land may be in better situations, although many regions cannot afford to divert land away from food production.

For third world countries, biodiesel sources that use marginal land could make more sense; e.g., pongam oiltree nuts grown along roads or jatropha grown along rail lines.

In tropical regions, such as Malaysia and Indonesia, plants that produce palm oil are being planted at a rapid pace to supply growing biodiesel demand in Europe and other markets. Scientists have shown that the removal of rainforest for palm plantations is not ecologically sound since the expansion of oil palm plantations poses a threat to natural rainforest and biodiversity.

It has been estimated in Germany that palm oil biodiesel has less than one third of the production costs of rapeseed biodiesel. The direct source of the energy content of biodiesel is solar energy captured by plants during photosynthesis. Regarding the positive energy balance of biodiesel:

> When straw was left in the field, biodiesel production was strongly energy positive, yielding 1 GJ biodiesel for every 0.561 GJ of energy input (a yield/cost ratio of 1.78).

> When straw was burned as fuel and oilseed rapemeal was used as a fertilizer, the yield/cost ratio for biodiesel production was even better (3.71). In other words, for every unit of energy input to produce biodiesel, the output was 3.71 units (the difference of 2.71 units would be from solar energy).

Economic Impact

Multiple economic studies have been performed regarding the economic impact of biodiesel production. One study, commissioned by the National Biodiesel Board, reported the 2011 production of biodiesel supported 39,027 jobs and more than $2.1 billion in household income. The growth in biodiesel also helps significantly increase GDP. In 2011, biodiesel created more than $3 billion in GDP. Judging by the continued growth in the Renewable Fuel Standard and the extension of the biodiesel tax incentive, the number of jobs can increase to 50,725, $2.7 billion in income, and reaching $5 billion in GDP by 2012 and 2013.

Energy Security

One of the main drivers for adoption of biodiesel is energy security. This means that a nation's dependence on oil is reduced, and substituted with use of locally available sources, such as coal, gas, or renewable sources. Thus a country can benefit from adoption of biofuels, without a reduction in greenhouse gas emissions. While the total energy balance is debated, it is clear that the dependence on oil is reduced. One example is the energy used to manufacture fertilizers, which could come from a variety of sources other than petroleum. The US National Renewable Energy Laboratory (NREL) states that energy security is the number one driving force behind the US biofuels programme, and a White House "Energy Security for the 21st Century" paper makes it clear that energy security is a major reason for promoting biodiesel. The former EU commission president, Jose Manuel Barroso, speaking at a recent EU biofuels conference, stressed that properly managed biofuels have the potential to reinforce the EU's security of supply through diversification of energy sources.

Global Biofuel Policies

Many countries around the world are involved in the growing use and production of biofuels, such as biodiesel, as an alternative energy source to fossil fuels and oil. To foster the biofuel industry, governments have implemented legislations and laws as incentives to reduce oil dependency and to increase the use of renewable energies. Many countries have their own independent policies regarding the taxation and rebate of biodiesel use, import, and production.

Canada

It was required by the Canadian Environmental Protection Act Bill C-33 that by the year 2010, gasoline contained 5% renewable content and that by 2013, diesel and heating oil contained 2% renewable content. The EcoENERGY for Biofuels Program subsidized the production of biodiesel, among other biofuels, via an incentive rate of CAN$0.20 per liter from 2008 to 2010. A decrease of $0.04 will be applied every year following, until the incentive rate reaches $0.06 in 2016. Individual provinces also have specific legislative measures in regards to biofuel use and production.

United States

The Volumetric Ethanol Excise Tax Credit (VEETC) was the main source of financial support for biofuels, but was scheduled to expire in 2010. Through this act, biodiesel production guaranteed a tax credit of US$1 per gallon produced from virgin oils, and $0.50 per gallon made from recycled oils. Currently soybean oil is being used to produce soybean biodiesel for many commercial purposes such as blending fuel for transportation sectors.

European Union

The European Union is the greatest producer of biodiesel, with France and Germany being the top producers. To increase the use of biodiesel, there exist policies requiring the blending of biodiesel into fuels, including penalties if those rates are not reached. In France, the goal was to reach 10% integration but plans for that stopped in 2010. As an incentive for the European Union countries to continue the production of the biofuel, there are tax rebates for specific quotas of biofuel produced. In Germany, the minimum percentage of biodiesel in transport diesel is set at 7% so called "B7".

Environmental Effects

The surge of interest in biodiesels has highlighted a number of environmental effects associated with its use. These potentially include reductions in greenhouse gas emissions, deforestation, pollution and the rate of biodegradation.

According to the EPA's Renewable Fuel Standards Program Regulatory Impact Analysis, released in February 2010, biodiesel from soy oil results, on average, in a 57% reduction in greenhouse gases compared to petroleum diesel, and biodiesel produced from waste grease results in an 86% reduction.

However, environmental organizations, for example, Rainforest Rescue and Greenpeace, criticize the cultivation of plants used for biodiesel production, e.g., oil palms, soybeans and sugar cane. They say the deforestation of rainforests exacerbates climate change and that sensitive ecosystems are destroyed to clear land for oil palm, soybean and sugar cane plantations. Moreover, that biofuels contribute to world hunger, seeing as arable land is no longer used for growing foods. The Environmental Protection Agency(EPA) published data in January 2012, showing that biofuels made from palm oil won't count towards the nation's renewable fuels mandate as they are not climate-friendly. Environmentalists welcome the conclusion because the growth of oil palm plantations has driven tropical deforestation, for example, in Indonesia and Malaysia.

Food, Land and Water Vs. Fuel

In some poor countries the rising price of vegetable oil is causing problems. Some propose that fuel only be made from non-edible vegetable oils such as camelina, jatropha or seashore mallow which can thrive on marginal agricultural land where many trees and crops will not grow, or would produce only low yields.

Others argue that the problem is more fundamental. Farmers may switch from producing food crops to producing biofuel crops to make more money, even if the new crops are not edible. The law of supply and demand predicts that if fewer farmers are producing food the price of food will rise. It may take some time, as farmers can take some time to change which things they are growing, but increasing demand for first generation biofuels is likely to result in price increases for many kinds of food. Some have pointed out that there are poor farmers and poor countries who are making more money because of the higher price of vegetable oil.

Biodiesel from sea algae would not necessarily displace terrestrial land currently used for food production and new algaculture jobs could be created.

Current Research

There is ongoing research into finding more suitable crops and improving oil yield. Other sources are possible including human fecal matter, with Ghana building its first "fecal sludge-fed biodiesel plant." Using the current yields, vast amounts of land and fresh water would be needed to produce enough oil to completely replace fossil fuel usage. It would require twice the land area of the US to be devoted to soybean production, or two-thirds to be devoted to rapeseed production, to meet current US heating and transportation needs.

Specially bred mustard varieties can produce reasonably high oil yields and are very useful in crop rotation with cereals, and have the added benefit that the meal leftover after the oil has been pressed out can act as an effective and biodegradable pesticide.

The NFESC, with Santa Barbara-based Biodiesel Industries is working to develop biodiesel technologies for the US navy and military, one of the largest diesel fuel users in the world.

A group of Spanish developers working for a company called Ecofasa announced a new biofuel made from trash. The fuel is created from general urban waste which is treated by bacteria to produce fatty acids, which can be used to make biodiesel.

Another approach that does not require the use of chemical for the production involves the use of genetically modified microbes.

Algal Biodiesel

From 1978 to 1996, the U.S. NREL experimented with using algae as a biodiesel source in the "Aquatic Species Program". A self-published article by Michael Briggs, at the UNH Biodiesel Group, offers estimates for the realistic replacement of all vehicular fuel with biodiesel by utilizing algae that have a natural oil content greater than 50%, which Briggs suggests can be grown on algae ponds at wastewater treatment plants. This oil-rich algae can then be extracted from the system and processed into biodiesel, with the dried remainder further reprocessed to create ethanol.

The production of algae to harvest oil for biodiesel has not yet been undertaken on a commercial scale, but feasibility studies have been conducted to arrive at the above yield estimate. In addition to its projected high yield, algaculture — unlike crop-based biofuels — does not entail a decrease in food production, since it requires neither farmland nor fresh water. Many companies are pursuing algae bio-reactors for various purposes, including scaling up biodiesel production to commercial levels.

Prof. Rodrigo E. Teixeira from the University of Alabama in Huntsville demonstrated the extraction of biodiesel lipids from wet algae using a simple and economical reaction in ionic liquids.

Pongamia

Millettia pinnata, also known as the Pongam Oiltree or Pongamia, is a leguminous, oilseed-bearing tree that has been identified as a candidate for non-edible vegetable oil production.

Pongamia plantations for biodiesel production have a two-fold environmental benefit. The trees both store carbon and produce fuel oil. Pongamia grows on marginal land not fit for food crops and does not require nitrate fertilizers. The oil producing tree has the highest yield of oil producing plant (approximately 40% by weight of the seed is oil) while growing in malnourished soils with high levels of salt. It is becoming a main focus in a number of biodiesel research organizations. The main advantages of Pongamia are a higher recovery and quality of oil than other crops and no direct competition with food crops. However, growth on marginal land can lead to lower oil yields which could cause competition with food crops for better soil.

Jatropha

Jatropha Biodiesel from DRDO, India.

Several groups in various sectors are conducting research on Jatropha curcas, a poisonous shrub-like tree that produces seeds considered by many to be a viable source of biodiesel feedstock oil. Much of this research focuses on improving the overall per acre oil yield of Jatropha through advancements in genetics, soil science, and horticultural practices.

SG Biofuels, a San Diego-based Jatropha developer, has used molecular breeding and biotechnology to produce elite hybrid seeds of Jatropha that show significant yield improvements over first generation varieties. SG Biofuels also claims that additional benefits have arisen from such strains, including improved flowering synchronicity, higher resistance to pests and disease, and increased cold weather tolerance.

Plant Research International, a department of the Wageningen University and Research Centre in the Netherlands, maintains an ongoing Jatropha Evaluation Project (JEP) that examines the feasibility of large scale Jatropha cultivation through field and laboratory experiments.

The Center for Sustainable Energy Farming (CfSEF) is a Los Angeles-based non-profit research organization dedicated to Jatropha research in the areas of plant science, agronomy, and horticulture. Successful exploration of these disciplines is projected to increase Jatropha farm production yields by 200-300% in the next ten years.

Fungi

A group at the Russian Academy of Sciences in Moscow published a paper in September 2008, stating that they had isolated large amounts of lipids from single-celled fungi and turned it into biodiesel in an economically efficient manner. More research on this fungal species; *Cunninghamella japonica*, and others, is likely to appear in the near future.

The recent discovery of a variant of the fungus *Gliocladium roseum* points toward the production of so-called myco-diesel from cellulose. This organism was recently discovered in the rainforests of northern Patagonia and has the unique capability of converting cellulose into medium length hydrocarbons typically found in diesel fuel.

Biodiesel from Used Coffee Grounds

Researchers at the University of Nevada, Reno, have successfully produced biodiesel from oil derived from used coffee grounds. Their analysis of the used grounds showed a 10% to 15% oil content (by weight). Once the oil was extracted, it underwent conventional processing into biodiesel. It is estimated that finished biodiesel could be produced for about one US dollar per gallon. Further, it was reported that "the technique is not difficult" and that "there is so much coffee around that several hundred million gallons of biodiesel could potentially be made annually." However, even if all the coffee grounds in the world were used to make fuel, the amount produced would be less than 1 percent of the diesel used in the United States annually. "It won't solve the world's energy problem," Dr. Misra said of his work.

Exotic Sources

Recently, alligator fat was identified as a source to produce biodiesel. Every year, about 15 million pounds of alligator fat are disposed of in landfills as a waste byproduct of the alligator meat and skin industry. Studies have shown that biodiesel produced from alligator fat is similar in composition to biodiesel created from soybeans, and is cheaper to refine since it is primarily a waste product.

Biodiesel to Hydrogen-cell Power

A microreactor has been developed to convert biodiesel into hydrogen steam to power fuel cells.

Steam reforming, also known as fossil fuel reforming is a process which produces hydrogen gas from hydrocarbon fuels, most notably biodiesel due to its efficiency. A **microreactor**, or reformer, is the processing device in which water vapour reacts with the liquid fuel under high temperature and pressure. Under temperatures ranging from 700 – 1100 °C, a nickel-based catalyst enables the production of carbon monoxide and hydrogen:

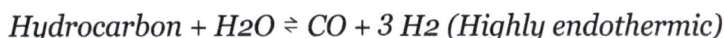

$$Hydrocarbon + H_2O \rightleftharpoons CO + 3\ H_2\ (Highly\ endothermic)$$

Furthermore, a higher yield of hydrogen gas can be harnessed by further oxidizing carbon monoxide to produce more hydrogen and carbon dioxide:

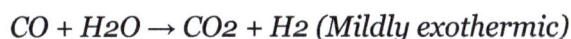

$$CO + H_2O \rightarrow CO_2 + H_2\ (Mildly\ exothermic)$$

Hydrogen Fuel Cells Background Information

Fuel cells operate similar to a battery in that electricity is harnessed from chemical reactions. The difference in fuel cells when compared to batteries is their ability to be powered by the constant flow of hydrogen found in the atmosphere. Furthermore, they produce only water as a by-product, and are virtually silent. The downside of hydrogen powered fuel cells is the high cost and dangers of storing highly combustible hydrogen under pressure.

One way new processors can overcome the dangers of transporting hydrogen is to produce it as necessary. The microreactors can be joined to create a system that heats the hydrocarbon under high pressure to generate hydrogen gas and carbon dioxide, a process called steam reforming. This produces up to 160 gallons of hydrogen/minute and gives the potential of powering hydrogen refueling stations, or even an on-board hydrogen fuel source for hydrogen cell vehicles. Implementation into cars would allow energy-rich fuels, such as biodiesel, to be transferred to kinetic energy while avoiding combustion and pollutant byproducts. The hand-sized square piece of metal contains microscopic channels with catalytic sites, which continuously convert biodiesel, and even its glycerol byproduct, to hydrogen.

Concerns

Engine Wear

Lubricity of fuel plays an important role in wear that occurs in an engine. An engine relies on its fuel to provide lubricity for the metal components that are constantly in contact with each other. Biodiesel is a much better lubricant compared with petroleum diesel due to the presence of esters. Tests have shown that the addition of a small amount of biodiesel to diesel can significantly increase the lubricity of the fuel in short term. However, over a longer period of time (2–4 years), studies show that biodiesel loses its lubricity. This could be because of enhanced corrosion over time due to oxidation of the unsaturated molecules or increased water content in biodiesel from moisture absorption.

Fuel Viscosity

One of the main concerns regarding biodiesel is its viscosity. The viscosity of diesel is 2.5–3.2 cSt at 40 °C and the viscosity of biodiesel made from soybean oil is between 4.2 and 4.6 cSt The viscosity of diesel must be high enough to provide sufficient lubrication for the engine parts but low enough to flow at operational temperature. High viscosity can plug the fuel filter and injection system in engines. Vegetable oil is composed of lipids with long chains of hydrocarbons, to reduce its viscosity the lipids are broken down into smaller molecules of esters. This is done by converting vegetable oil and animal fats into alkyl esters using transesterification to reduce their viscosity

Nevertheless, biodiesel viscosity remains higher than that of diesel, and the engine may not be able to use the fuel at low temperatures due to the slow flow through the fuel filter.

Engine Performance

Biodiesel has higher brake-specific fuel consumption compared to diesel, which means more biodiesel fuel consumption is required for the same torque. However, B20 biodiesel blend has been found to provide maximum increase in thermal efficiency, lowest brake-specific energy consumption, and lower harmful emissions. The engine performance depends on the properties of the fuel, as well as on combustion, injector pressure and many other factors. Since there are various blends of biodiesel, that may account for the contradicting reports in regards engine performance.

Environmental Impact of Biodiesel

The environmental impact of biodiesel is diverse.

Greenhouse Gas Emissions

Carbon Intensity of Biodiesel Production (US Soy Oil Esterfied in the UK)

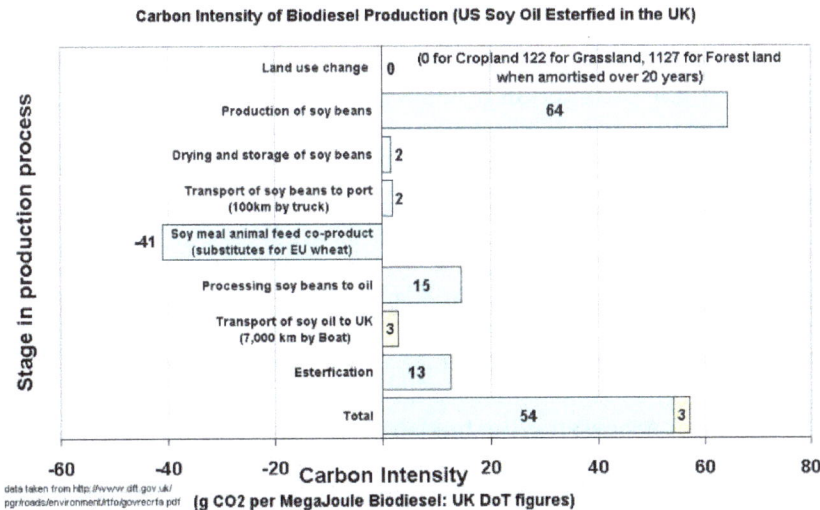

Calculation of Carbon Intensity of Soy biodiesel grown in the US and burnt in the UK, using figures calculated by the UK government for the purposes of the Renewable transport fuel obligation.

An often mentioned incentive for using biodiesel is its capacity to lower greenhouse gas emissions compared to those of fossil fuels. Whether this is true or not depends on many factors. Especially the effects from land use change have potential to cause even more emissions than what would be caused by using fossil fuels alone.

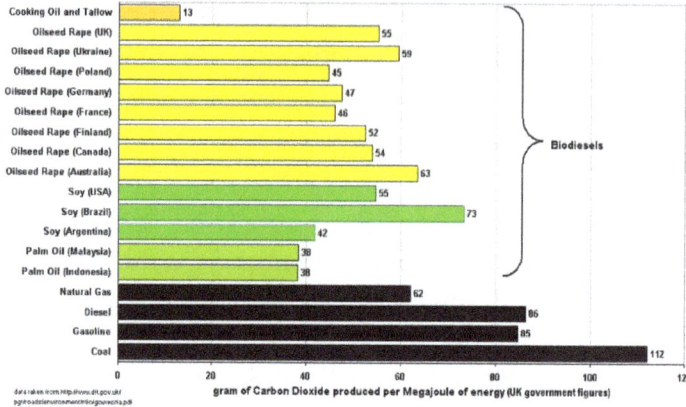

Graph of UK figures for the Carbon Intensity of Biodiesels and fossil fuels. This graph assumes that all biodiesel is used in its country of origin. It also assumes that the diesel is produced from pre-existing croplands rather than by changing land use

Carbon dioxide is one of the major greenhouse gases. Although the burning of biodiesel produces carbon dioxide emissions similar to those from ordinary fossil fuels, the plant feedstock used in the production absorbs carbon dioxide from the atmosphere when it grows. Plants absorb carbon dioxide through a process known as photosynthesis which allows it to store energy from sunlight in the form of sugars and starches. After the biomass is converted into biodiesel and burned as fuel the energy and carbon is released again. Some of that energy can be used to power an engine while the carbon dioxide is released back into the atmosphere.

When considering the total amount of greenhouse gas emissions it is therefore important to consider the whole production process and what indirect effects such production might cause. The effect on carbon dioxide emissions is highly dependent on production methods and the type of feedstock used. Calculating the carbon intensity of biofuels is a complex and inexact process, and is highly dependent on the assumptions made in the calculation. A calculation usually includes:

- Emissions from growing the feedstock (e.g. Petrochemicals used in fertilizers)

- Emissions from transporting the feedstock to the factory

- Emissions from processing the feedstock into biodiesel

Other factors can be very significant but are sometimes not considered. These include:

- Emissions from the change in land use of the area where the fuel feedstock is grown.

- Emissions from transportation of the biodiesel from the factory to its point of use

- The efficiency of the biodiesel compared with standard diesel

- The amount of Carbon Dioxide produced at the tail pipe. (Biodiesel can produce 4.7% more)

- The benefits due to the production of useful by-products, such as cattle feed or glycerine

If land use change is not considered and assuming today's production methods, biodiesel from rapeseed and sunflower oil produce 45%-65% lower greenhouse gas emissions than petrodiesel. However, there is ongoing research to improve the efficiency of the production process. Biodiesel produced from used cooking oil or other waste fat could reduce CO_2 emissions by as much as 85%. As long as the feedstock is grown on existing cropland, land use change has little or no effect on greenhouse gas emissions. However, there is concern that increased feedstock production directly affects the rate of deforestation. Such clearcutting cause carbon stored in the forest, soil and peat layers to be released. The amount of greenhouse gas emissions from deforestation is so large that the benefits from lower emissions (caused by biodiesel use alone) would be negligible for hundreds of years. Biofuel produced from feedstock such as palm oil could therefore cause much higher carbon dioxide emissions than some types of fossil fuels.

Pollution

In the United States, biodiesel is the only alternative fuel to have successfully completed the Health Effects Testing requirements (Tier I and Tier II) of the Clean Air Act (1990).

Biodiesel can reduce the direct tailpipe-emission of particulates, small particles of solid combustion products, on vehicles with particulate filters by as much as 20 percent compared with low-sulfur (< 50 ppm) diesel. Particulate emissions as the result of production are reduced by around 50 percent compared with fossil-sourced diesel. (Beer et al., 2004). Biodiesel has a higher cetane rating than petrodiesel, which can improve performance and clean up emissions compared to crude petro-diesel (with cetane lower than 40). Biodiesel contains fewer aromatic hydrocarbons: benzofluoranthene: 56% reduction; Benzopyrenes: 71% reduction.

Biodegradation

A University of Idaho study compared biodegradation rates of biodiesel, neat vegetable oils, biodiesel and petroleum diesel blends, and neat 2-D diesel fuel. Using low concentrations of the product to be degraded (10 ppm) in nutrient and sewage sludge amended solutions, they demonstrated that biodiesel degraded at the same rate as a dextrose control and 5 times as quickly as petroleum diesel over a period of 28 days, and that biodiesel blends doubled the rate of petroleum diesel degradation through co-metabolism. The same study examined soil degradation using 10 000 ppm of biodiesel and petroleum diesel, and found biodiesel degraded at twice the rate of petroleum diesel in soil. In all cases, it was determined biodiesel also degraded more completely than petroleum diesel, which produced poorly degradable undetermined intermediates. Toxicity studies for the same project demonstrated no mortalities and few toxic effects on rats and rabbits with up to 5000 mg/kg of biodiesel. Petroleum diesel showed no mortalities at the same concentration either, however toxic effects such as hair loss and urinary discolouring were noted with concentrations of >2000 mg/l in rabbits.:

Biodegradation in Aquatic Environments

As biodiesel becomes more widely used, it is important to consider how consumption affects water quality and aquatic ecosystems. Research examining the biodegradability of different biodiesel fuels found that all of the biofuels studied (including Neat Rapeseed oil, Neat Soybean oil, and their modified ester products) were "readily biodegradable" compounds, and had a relatively high biodegradation rate in water. Additionally, the presence of biodiesel can increase the rate of diesel biodegradation via co-metabolism. As the ratio of biodiesel is increased in biodiesel/diesel mixtures, the faster the diesel is degraded. Another study using controlled experimental conditions also showed that fatty acid methyl esters, the primary molecules in biodiesel, degraded much faster than petroleum diesel in sea water.

Carbonyl Emissions

When considering the emissions from fossil fuel and biofuel use, research typically focuses on major pollutants such as hydrocarbons. It is generally recognized that using biodiesel in place of diesel results in a substantial reduction in regulated gas emissions, but there has been a lack of information in research literature about the non-regulated compounds which also play a role in air pollution. One study focused on the emissions of non-criteria carbonyl compounds from the burning of pure diesel and biodiesel blends in heavy-duty diesel engines. The results found that carbonyl emissions of formaldehyde, acetaldehyde, acrolein, acetone, propionaldehyde and butyraldehyde, were higher in biodiesel mixtures than emissions from pure diesel. Biodiesel use results in higher carbonyl emissions but lower total hydrocarbon emissions, which may be better as an alternative fuel source. Other studies have been done which conflict with these results, but comparisons are difficult to make due to various factors that differ between studies (such as types of fuel and engines used). In a paper which compared 12 research articles on carbonyl emissions from biodiesel fuel use, it found that 8 of the papers reported increased carbonyl compound emissions while 4 showed the opposite. This is evidence that there is still much research required on these compounds.

Bioenergy with Carbon Capture and Storage

Bio-energy with carbon capture and storage produces negative carbon dioxide emissions by combining the use of bioenergy with geologic carbon capture and storage. This chapter will provide the readers with an integrated understanding of carbon capture and storage.

Bio-energy with carbon capture and storage (BECCS) is a future greenhouse gas mitigation technology which produces negative carbon dioxide emissions by combining bioenergy (energy from biomass) use with geologic carbon capture and storage. The concept of BECCS is drawn from the integration of trees and crops, which extract carbon dioxide (CO_2) from the atmosphere as they grow, the use of this biomass in processing industries or power plants, and the application of carbon capture and storage via CO_2 injection into geological formations. There are other non-BECCS forms of carbon dioxide removal and storage that include technologies such as biochar, carbon dioxide air capture and biomass burial.

According to a recent Biorecro report, there is 550 000 tonnes CO_2/year in total BECCS capacity currently operating, divided between three different facilities (as of January 2012).

In the IPCC Fourth Assessment Report by the Intergovernmental Panel on Climate Change (IPCC). BECCS was indicated as a key technology for reaching low carbon dioxide atmospheric concentration targets. The negative emissions that can be produced by BECCS has been estimated by the Royal Society to be equivalent to a 50 to 150 ppm decrease in global atmospheric carbon dioxide concentrations and according to the International Energy Agency, the BLUE map climate change mitigation scenario calls for more than 2 gigatonnes of negative CO_2 emissions per year with BECCS in 2050. According to Stanford University, 10 gigatonnes is achievable by this date.

The Imperial College London, the UK Met Office Hadley Centre for Climate Prediction and Research, the Tyndall Centre for Climate Change Research, the Walker Institute for Climate System Research, and the Grantham Institute for Climate Change issued a joint report on carbon dioxide removal technologies as part of the *AVOID: Avoiding dangerous climate change* research program, stating that "Overall, of the technologies studied in this report, BECCS has the greatest maturity and there are no major practical barriers to its introduction into today's energy system. The presence of a primary product will support early deployment."

According to the OECD, "Achieving lower concentration targets (450 ppm) depends significantly on the use of BECCS".

Negative Emission

The main appeal of BECCS is in its ability to result in negative emissions of CO_2. The capture of carbon dioxide from bioenergy sources effectively removes CO_2 from the atmosphere.

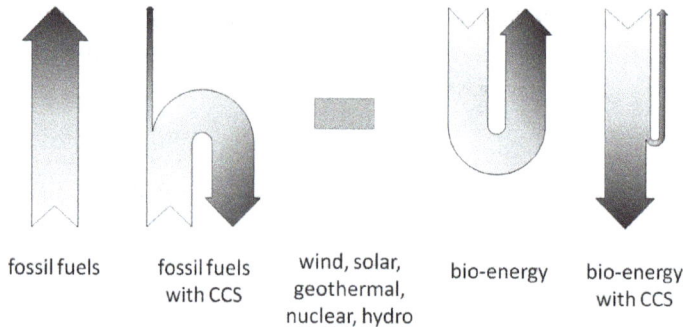

| fossil fuels | fossil fuels with CCS | wind, solar, geothermal, nuclear, hydro | bio-energy | bio-energy with CCS |

Carbon flow schematic for different energy systems.

Bio-energy is derived from biomass which is a renewable energy source and serves as a carbon sink during its growth. During industrial processes, the biomass combusted or processed re-releases the CO_2 into the atmosphere. The process thus results in a net zero emission of CO_2, though this may be positively or negatively altered depending on the carbon emissions associated with biomass growth, transport and processing. Carbon capture and storage (CCS) technology serves to intercept the release of CO_2 into the atmosphere and redirect it into geological storage locations. CO_2 with a biomass origin is not only released from biomass fuelled power plants, but also during the production of pulp used to make paper and in the production of biofuels such as biogas and bioethanol. The BECCS technology can also be employed on such industrial processes.

It is argued that through the BECCS technology, carbon dioxide is trapped in geologic formations for very long periods of time, whereas for example a tree only stores its carbon during its lifetime. In its report on the CCS technology, IPCC projects that more than 99% of carbon dioxide which is stored through geologic sequestration is likely to stay in place for more than 1000 years. While other types of carbon sinks such as the ocean, trees and soil may involve the risk of negative feedback loops at increased temperatures, BECCS technology is likely to provide a better permanence by storing CO_2 in geological formations.

The amount of CO_2 that has been released to date is believed to be too much to be able to be absorbed by conventional sinks such as trees and soil in order to reach low emission targets. In addition to the presently accumulated emissions, there will be significant additional emissions during this century, even in the most ambitious low-emission scenarios. BECCS has therefore been suggested as a technology to reverse the emission trend and create a global system of net negative emissions. This implies that the emissions would not only be zero, but negative, so that not only the emissions, but the absolute amount of CO_2 in the atmosphere would be reduced.

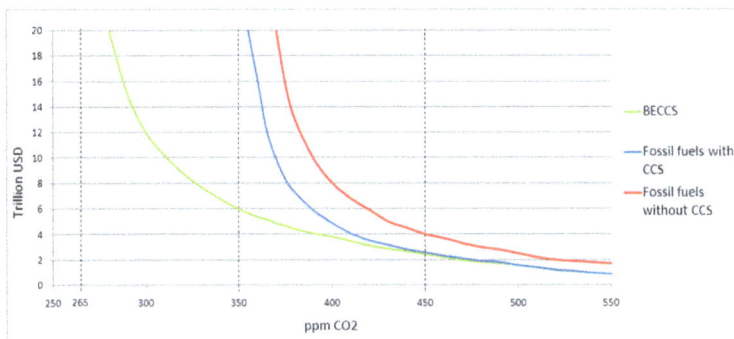

Projected cost to reach the respective 350ppm and 450ppm target scenarios by 2100. 265ppm indicates the pre-industrial atmospheric CO_2 level.

Application

Source	CO₂ Source	Sector
Electrical power plants	Combustion of biomass or biofuel in steam or gas powered generators releases CO_2 as a by-product	Energy
Heat power plants	Combustion of biofuel for heat generation releases CO_2 as a by-product. Usually used for district heating	Energy
Pulp and paper mills	• CO_2 produced in recovery boilers • CO_2 produced in lime kilns • For gasification technologies, CO_2 is produced during the gasification of black liquor and biomass such as the tree bark and woody. • Huge amounts of CO_2 are also released by the combustion of syngas, a product of gasification, in the combined cycle process.	Industry
Ethanol production	Fermentation of biomass such as sugarcane, wheat or corn releases CO_2 as a by-product	Industry
Biogas production	In the biogas upgrading process, CO_2 is separated from the methane to produce a higher quality gas	Industry

Technology

The main technology for CO_2 capture from biotic sources generally employs the same technology as carbon dioxide capture from conventional fossil fuel sources. Broadly, three different types of technologies exist: post-combustion, pre-combustion, and oxy-fuel combustion.

However, on the critical subject of *"upstream large-scale provision of the biomass"* (IPCC WG3 Summary for Policymakers) cultivation/refinery technology, marine based BECCS (i.e. MBECS) operations, within the sub-tropical convergence zones (STCZ), offers a unique combination of stable oceanic environmental conditions which can physically accommodate vast MBECS operations while avoiding the long list of limiting factors found within terrestrial BECCS. The STCZs are marine deserts with no practical levels of biological activity within the surface or nutricline waters and which also offers vast amounts of renewable energy for cultivation/processing of the biomass.

The raw nutrients found within the nutricline can be used to support vast scale biomass production with only mineral importation. The cultivation of both micro and macro algal species can be augmented with terrestrial species such as halophytes (for oil) and bamboo (for Cellulosic ethanol) grown through aquaponic sub-systems. The use of non-photosynthesis dependent cultivation methods expands micro algal cultivation into all three dimensions. Farming *down* the water column to 50+ meters is possible. Also, a broad spectrum of critical (non-fuel) commodities, such as food/feed/fertilizer/plastic and vast volumes of freshwater production, can economically subsidize MBECS biofuel production at sub fossil fuel prices.

Based upon reasonable assumptions concerning biofuel production output for 1 km2 under the MBECS scenario, the total global replacement of fossil fuels can be achieved at <1.5M km2 of MBECS operations. Global replacement of fossil fuels is achievable within 20 years, while within that same time frame, all energy importing nations can achieve energy independence through involvement in the IMBECS Protocol.

The coordination of an internationally collaborative MBECS effort is the focus of the Intergovernmental Marine Bio-Energy and Carbon Sequestration (IMBECS) Protocol (developed by Michael Hayes).

Cost

The sustainable technical potential for net negative emissions with BECCS has been estimated to 10 Gt of CO_2 equivalent annually, with an economic potential of up to 3.5 Gt of CO_2 equivalent annually at a cost of less than 50 €/tonne, and up to 3.9 Gt of CO_2 equivalent annually at a cost of less than 100 €/tonne.

Policy

Based on the current Kyoto Protocol agreement, carbon capture and storage projects are not applicable as an emission reduction tool to be used for the Clean Development Mechanism (CDM) or for Joint Implementation (JI) projects. Recognising CCS technologies as an emission reduction tool is vital for the implementation of such plants as there is no other financial motivation for the implementation of such systems. There has been growing support to have fossil CCS and BECCS included in the protocol. Accounting studies on how this can be implemented, including BECCS, have also been done.

Techno-economics of BECCS and The TESBiC Project

The largest and most detailed techno-economic assessment of BECCS was carried out by cmcl innovations and the TESBiC group (Techno-Economic Study of Biomass to CCS) in 2012. This project recommended the most promising set of biomass fuelled power generation technologies coupled with carbon capture and storage (CCS). The project outcomes lead to a detailed "biomass CCS roadmap" for the U.K..

Environmental Considerations

Some of the environmental considerations and other concerns about the widespread implementation of BECCS are similar to those of CCS. However, much of the critique towards CCS is that it may strengthen the dependency on depletable fossil fuels and environmentally invasive coal mining. This is not the case with BECCS, as it relies on renewable biomass. There are however other considerations which involve BECCS and these concerns are related to the possible increased use of biofuels.

Biomass production is subject to a range of sustainability constraints, such as: scarcity of arable land and fresh water, loss of biodiversity, competition with food production, deforestation and scarcity of phosphorus. It is important to make sure that biomass is used in a way that maximizes both energy and climate benefits. There has been criticism to some suggested BECCS deployment scenarios, where there would be a very heavy reliance on increased biomass input.

These systems may have other negative side effects. There is however presently no need to expand the use of biofuels in energy or industry applications to allow for BECCS deployment. There is already today considerable emissions from point sources of biomass derived CO_2, which could be utilized for BECCS. Though, in possible future bio-energy system upscaling scenarios, this may be an important consideration.

The BECCS process allows CO_2 to be collected and stored directly from the atmosphere, rather than from a fossil source. This implies that any eventual emissions from storage may be recollected and restored simply by reiterating the BECCS-process. This is not possible with CCS alone, as CO_2 emitted to the atmosphere cannot be restored by burning more fossil fuel with CCS.

References

- Gaffron, Hans (1942-11-20), "Reduction of Carbon Dioxide Coupled with the Oxyhydrogen Reaction in Algae" (PDF), The Journal of General Physiology: 241–267, retrieved 2015-12-10 .

- Future Initiatives, Ocean Thermal Energy Corporation, archived from the original (pdf) on 2015-02-15, retrieved 2015-12-10 .

- "Global Technology Roadmap for CCS in Industry Biomass-based industrial CO2 sources: biofuels production with CCS" (PDF). ECN. 2011. Retrieved 2012-01-20.

- "First U.S. large demonstration-scale injection of CO2 from a biofuel production facility begins". Retrieved 20 January 2012.

- "The Potential for the Deployment of Negative Emissions Technologies in the UK" (PDF). Grantham Institute for Climate Change, Imperial College. 2010. Retrieved 2012-01-16.

- "OECD Environmental Outlook to 2050, Climate Change Chapter, pre-release version" (PDF). OECD. 2011. Retrieved 2012-01-16.

Various Aspects of Bioenergy

The aspects of bioenergy are sustainable development and sustainability. Sustainable development is the procedure of meeting human development goals; this is achieved by sustaining natural resources and protecting the environment. The section strategically encompasses and incorporates the major aspects of bioenergy, providing the reader with a complete understanding on the respective subject.

Sustainable Development

Sustainable development is a process for meeting human development goals while sustaining the ability of natural systems to continue to provide the natural resources and ecosystem services upon which the economy and society depend. While the modern concept of sustainable development is derived most strongly from the 1987 Brundtland Report, it is rooted in earlier ideas about sustainable forest management and twentieth century environmental concerns. As the concept developed, it has shifted to focus more on economic development, social development and environmental protection.

Wind powers 5 MW wind turbines on a wind farm 28 km off the coast of Belgium.

Sustainable development is the organizing principle for sustaining finite resources necessary to provide for the needs of future generations of life on the planet. It is a process that envisions a desirable future state for human societies in which living conditions and resource-use continue to meet human needs without undermining the "integrity, stability and beauty" of natural biotic systems.

History

Sustainability can be defined as the practice of maintaining processes of productivity indefinitely—natural or human made—by replacing resources used with resources of equal or greater value without degrading or endangering natural biotic systems. Sustainable development ties together concern for the carrying capacity of natural systems with the social, political, and economic challenges faced by humanity. Sustainability science is the study of the concepts of sustainable development and environmental science. There is an additional focus on the present generations' responsibility to regenerate, maintain and improve planetary resources for use by future generations.

The Blue Marble, photographed from Apollo 17 in 1972, quickly became an icon of environmental conservation.

Sustainable development has its roots in ideas about sustainable forest management which were developed in Europe during the seventeenth and eighteenth centuries. In response to a growing aware of the depletion of timber resources in England, John Evelyn argued that "sowing and planting of trees had to be regarded as a national duty of every landowner, in order to stop the destructive over-exploitation of natural resources" in his 1662 essay *Sylva*. In 1713 Hans Carl von Carlowitz, a senior mining administrator in the service of Elector Frederick Augustus I of Saxony published *Sylvicultura oeconomica*, a 400-page work on forestry. Building upon the ideas of Evelyn and French minister Jean-Baptiste Colbert, von Carlowitz developed the concept of managing forests for sustained yield. His work influenced others, including Alexander von Humboldt and Georg Ludwig Hartig, leading in turn to the development of a science of forestry. This in term influenced people like Gifford Pinchot, first head of the US Forest Service, whose approach to forest management was driven by the idea of wise use of resources, and Aldo Leopold whose land ethic was influential in the development of the environmental movement in the 1960s.

Following the publication of Rachel Carson's *Silent Spring* in 1962, the developing environmental movement drew attention to the relationship between economic growth and development and environmental degradation. Kenneth E. Boulding in his influential 1966 essay *The Economics of the Coming Spaceship Earth* identified the need for the economic system to fit itself to the ecological system with its limited pools of resources. One of the first uses of the term sustainable in the contemporary sense was by the Club of Rome in 1972 in its classic report on the *Limits to Growth*, written by a group of scientists led by Dennis and Donella Meadows of the Massachusetts Institute of Technology. Describing the desirable "state of global equilibrium", the authors wrote: "We are searching for a model output that represents a world system that is sustainable without sudden and uncontrolled collapse and capable of satisfying the basic material requirements of all of its people."

In 1980 the International Union for the Conservation of Nature published a world conservation strategy that included one of the first references to sustainable development as a global priority and introduced the term "sustainable development". Two years later, the United Nations World Charter for Nature raised five principles of conservation by which human conduct affecting nature is to be guided and judged. In 1987 the United Nations World Commission on Environment and Development released the report *Our Common Future*, commonly called the Brundtland Report. The report included what is now one of the most widely recognised definitions of sustainable development.

> Sustainable development is development that meets the needs of the present without compromising the ability of future generations to meet their own needs. It contains within it two key concepts:
> · The concept of 'needs', in particular, the essential needs of the world's poor, to which overriding priority should be given; and· The idea of limitations imposed by the state of technology and social organization on the environment's ability to meet present and future needs.
>
> — World Commission on Environment and Development, *Our Common Future* (1987)

Since the Brundtland Report, the concept of sustainable development has developed beyond the initial intergenerational framework to focus more on the goal of "socially inclusive and environmentally sustainable economic growth". In 1992, the UN Conference on Environment and Development published the Earth Charter, which outlines the building of a just, sustainable, and peaceful global society in the 21st century. The action plan Agenda 21 for sustainable development identified information, integration, and participation as key building blocks to help countries achieve development that recognises these interdependent pillars. It emphasises that in sustainable development everyone is a user and provider of information. It stresses the need to change from old sector-centered ways of doing business to new approaches that involve cross-sectoral co-ordination and the integration of environmental and social concerns into all development processes. Furthermore, Agenda 21 emphasises that broad public participation in decision making is a fundamental prerequisite for achieving sustainable development.

Under the principles of the United Nations Charter the Millennium Declaration identified principles and treaties on sustainable development, including economic development, social development and environmental protection. Broadly defined, sustainable development is a systems approach to growth and development and to manage natural, produced, and social capital for the welfare of their own and future generations. The term sustainable development as used by the United Nations incorporates both issues associated with land development and broader issues of human development such as education, public health, and standard of living.

A 2013 study concluded that sustainability reporting should be reframed through the lens of four interconnected domains: ecology, economics, politics and culture.

The Sustainable Development Goals (SDGs)

On September 2015, the United Nations General Assembly formally adopted the "universal, integrated and transformative" 2030 Agenda for Sustainable Development, a set of 17 Sustainable Development Goals (SDGs).

These icons represent the 17 headline SDGs. There are 169 targets under the goals.

The goals are to be implemented and achieved in every country from the year 2016 to 2030.

Dimensions

Sustainable development has been described in terms of three dimensions, domains or pillars. In the three-dimension model, these are seen as "economic, environmental and social" or "ecology, economy and equity"; this has been expanded by some authors to include a fourth pillar of culture, institutions or governance.

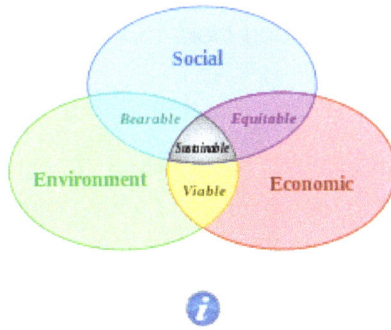

Scheme of sustainable development:
at the confluence of three constituent parts. (2006)

Environmental

The ecological stability of human settlements is part of the relationship between humans and their natural, social and built environments. Also termed human ecology, this broadens the focus of sustainable development to include the domain of human health. Fundamental human needs such as the availability and quality of air, water, food and shelter are also the ecological foundations for sustainable development; addressing public health risk through investments in ecosystem services can be a powerful and transformative force for sustainable development which, in this sense, extends to all species.

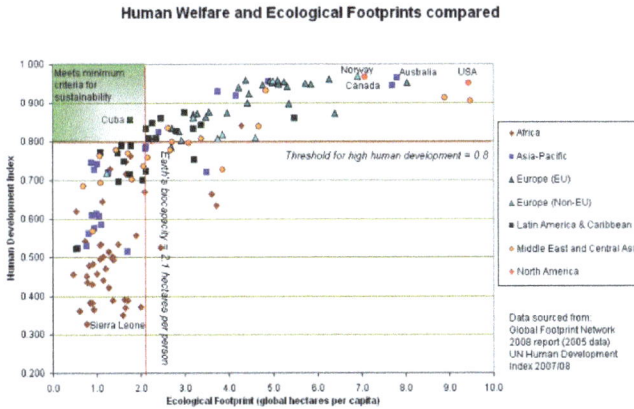

Relationship between ecological footprint and Human Development Index (HDI)

Environmental sustainability concerns the natural environment and how it endures and remains diverse and productive. Since natural resources are derived from the environment, the state of air, water, and the climate are of particular concern. The IPCC Fifth Assessment Report outlines current knowledge about scientific, technical and socio-economic information concerning climate change, and lists options for adaptation and mitigation. Environmental sustainability requires society to design activities to meet human needs while preserving the life support systems of the planet. This, for example, entails using water sustainably, utilizing renewable energy, and sustainable material supplies (e.g. harvesting wood from forests at a rate that maintains the biomass and biodiversity).

An unsustainable situation occurs when natural capital (the sum total of nature's resources) is used up faster than it can be replenished. Sustainability requires that human activity only uses nature's resources at a rate at which they can be replenished naturally. Inherently the concept of sustainable development is intertwined with the concept of carrying capacity. Theoretically, the long-term result of environmental degradation is the inability to sustain human life. Such degradation on a global scale should imply an increase in human death rate until population falls to what the degraded environment can support. If the degradation continues beyond a certain tipping point or critical threshold it would lead to eventual extinction for humanity.

Consumption of non-renewable resources	State of environment	Sustainability
More than nature's ability to replenish	Environmental degradation	Not sustainable
Equal to nature's ability to replenish	Environmental equilibrium	Steady state economy
Less than nature's ability to replenish	Environmental renewal	Environmentally sustainable

Integral elements for a sustainable development are research and innovation activities. A telling example is the European environmental research and innovation policy, which aims at defining and implementing a transformative agenda to greening the economy and the society as a whole so to achieve a truly sustainable development. Research and innovation in Europe is financially supported by the programme Horizon 2020, which is also open to participation worldwide. A promising direction towards sustainable development is to design systems that are flexible and reversible.

Pollution of the public resources is really not a different action, it just is a reverse tragedy of the commons, in that instead of taking something out, something is put into the commons. When the costs of polluting the commons are not calculated into the cost of the items consumed, then it becomes only natural to pollute, as the cost of pollution is external to the cost of the goods produced and the cost of cleaning the waste before it is discharged exceeds the cost of releasing the waste directly into the commons. So, the only way to solve this problem is by protecting the ecology of the commons by making it, through taxes or fines, more costly to release the waste directly into the commons than would be the cost of cleaning the waste before discharge.

So, one can try to appeal to the ethics of the situation by doing the right thing as an individual, but in the absence of any direct consequences, the individual will tend to do what is best for the person and not what is best for the common good of the public. Once again, this issue needs to be addressed. Because, left unaddressed, the development of the commonly owned property will become impossible to achieve in a sustainable way. So, this topic is central to the understanding of creating a sustainable situation from the management of the public resources that are used for personal use.

Agriculture

Sustainable agriculture consists of environmentally-friendly methods of farming that allow the production of crops or livestock without damage to human or natural systems. It involves preventing adverse effects to soil, water, biodiversity, surrounding or downstream resources—as well as to those working or living on the farm or in neighboring areas. The concept of sustainable agriculture extends intergenerationally, passing on a conserved or improved natural resource, biotic, and economic base rather than one which has been depleted or polluted. Elements of sustainable agriculture include permaculture, agroforestry, mixed farming, multiple cropping, and crop rotation.

Numerous sustainability standards and certification systems exist, including organic certification, Rainforest Alliance, Fair Trade, UTZ Certified, Bird Friendly, and the Common Code for the Coffee Community (4C).

Economics

It has been suggested that because of rural poverty and overexploitation, environmental resources should be treated as important economic assets, called natural capital. Economic development has traditionally required a growth in the gross domestic product. This model of unlimited personal and GDP growth may be over. Sustainable development may involve improvements in the quality of life for many but may necessitate a decrease in resource consumption. According to ecological economist Malte Faber, ecological economics is defined by its focus on nature, justice, and time. Issues of intergenerational equity, irreversibility of environmental change, uncertainty of long-term outcomes, and sustainable development guide ecological economic analysis and valuation.

A sewage treatment plant that uses solar energy, located at Santuari de Lluc monastery, Majorca.

As early as the 1970s, the concept of sustainability was used to describe an economy "in equilibrium with basic ecological support systems." Scientists in many fields have highlighted *The Limits to Growth*, and economists have presented alternatives, for example a 'steady-state economy'; to address concerns over the impacts of expanding human development on the planet. In 1987 the economist Edward Barbier published the study

The Concept of Sustainable Economic Development, where he recognised that goals of environmental conservation and economic development are not conflicting and can be reinforcing each other.

A World Bank study from 1999 concluded that based on the theory of genuine savings, policymakers have many possible interventions to increase sustainability, in macro-economics or purely environmental. A study from 2001 noted that efficient policies for renewable energy and pollution are compatible with increasing human welfare, eventually reaching a golden-rule steady state. The study, *Interpreting Sustainability in Economic Terms*, found three pillars of sustainable development, interlinkage, inter-generational equity, and dynamic efficiency.

But Gilbert Rist points out that the World Bank has twisted the notion of sustainable development to prove that economic development need not be deterred in the interest of preserving the ecosystem. He writes: "From this angle, 'sustainable development' looks like a cover-up operation. ... The thing that is meant to be sustained is really 'development', not the tolerance capacity of the ecosystem or of human societies."

The World Bank, a leading producer of environmental knowledge, continues to advocate the win-win prospects for economic growth and ecological stability even as its economists express their doubts. Herman Daly, an economist for the Bank from 1988 to 1994, writes:

When authors of *WDR* '92 [the highly influential 1992 *World Development Report* that featured the environment] were drafting the report, they called me asking for examples of "win-win" strategies in my work. What could I say? None exists in that pure form; there are trade-offs, not "win-wins." But they want to see a world of "win-wins" based on articles of faith, not fact. I wanted to contribute because *WDRs* are important in the Bank, [because] task managers read [them] to find philosophical justification for their latest round of projects. But they did not want to hear about how things really are, or what I find in my work..."

A meta review in 2002 looked at environmental and economic valuations and found a lack of "sustainability policies". A study in 2004 asked if we consume too much. A study concluded in 2007 that knowledge, manufactured and human capital (health and education) has not compensated for the degradation of natural capital in many parts of the world. It has been suggested that intergenerational equity can be incorporated into a sustainable development and decision making, as has become common in economic valuations of climate economics. A meta review in 2009 identified conditions for a strong case to act on climate change, and called for more work to fully account of the relevant economics and how it affects human welfare. According to free-market environmentalist John Baden "the improvement of environment quality depends on the market economy and the existence of legitimate and protected property rights." They enable the effective practice of personal responsibility and the development of mechanisms to protect the environment. The State can in this context "create conditions which encourage the people to save the environment."

Environmental Economics

The total environment includes not just the biosphere of earth, air, and water, but also human interactions with these things, with nature, and what humans have created as their surroundings.

As countries around the world continue to advance economically, they put a strain on the ability of the natural environment to absorb the high level of pollutants that are created as a part of this economic growth. Therefore, solutions need to be found so that the economies of the world can continue to grow, but not at the expense of the public good. In the world of economics the amount of environmental quality must be considered as limited in supply and therefore is treated as a scarce resource. This is a resource to be protected and the only real efficient way to do it in a market economy is to look at the overall situation of pollution from a benefit-cost perspective. It then becomes essentially an allocation of resources, based on an evaluation of the expected course of action and the consequences of this action, when compared to an alternative course of action that might allocate the limited resources in a different way.

Benefit-cost analysis basically can look at several ways of solving a problem and then assigning the best route for a solution, based on the set of consequences that would result from the further development of the individual courses of action, and then choosing the course of action that results in the least amount of damage to the expected outcome for the environmental quality that remains after that development or process takes place. Further complicating this analysis are the interrelationships of the various parts of the environment that might be impacted by the chosen course of action. Sometimes it is almost impossible to predict the various outcomes of a course of action, due to the unexpected consequences and the amount of unknowns that are not accounted for in the benefit-cost analysis.

Energy

Sustainable energy is clean and can be used over a long period of time. Unlike fossil fuels that most countries are using, renewable energy only produces little or even no pollution. The most common types of renewable energy in US are hydroelectric, solar and wind energy. Solar energy is commonly used on public parking meters, street lights and the roof of buildings. Wind power has expanded quickly, it's share of worldwide electricity usage at the end of 2014 was 3.1%. Most of California's fossil fuel infrastructures are sited in or near low-income communities, and have traditionally suffered the most from California's fossil fuel energy system. These communities are historically left out during the decision-making process, and often end up with dirty power plants and other dirty energy projects that poison the air and harm the area. These toxicants are major contributors to health problems in the communities. As renewable energy becomes more common, fossil fuel infrastructures are replaced by renewables, providing better social equity to these communities. Overall, and in the long run, sustainable development in the field of energy is also deemed to contribute to economic sustainability and national security of communities, thus being increasingly encouraged through investment policies.

Manufacturing

Technology

One of the core concepts in sustainable development is that technology can be used to assist people meet their developmental needs. Technology to meet these sustainable development needs is often referred to as appropriate technology, which is an ideological movement (and its manifestations) originally articulated as intermediate technology by the economist E. F. Schumacher in his influential work, *Small is Beautiful*. and now covers a wide range of technologies. Both Schumacher and many modern-day proponents of appropriate technology also emphasise the technology as people-centered. Today appropriate technology is often developed using open source principles, which have led to open-source appropriate technology (OSAT) and thus many of the plans of the technology can be freely found on the Internet. OSAT has been proposed as a new model of enabling innovation for sustainable development.

Transport

Transportation is a large contributor to greenhouse gas emissions. It is said that one-third of all gasses produced are due to transportation. Motorized transport also releases exhaust fumes that contain particulate matter which is hazardous to human health and a contributor to climate change.

Sustainable transport has many social and economic benefits that can accelerate local sustainable development. According to a series of reports by the Low Emission Development Strategies Global Partnership (LEDS GP), sustainable transport can help create jobs, improve commuter safety through investment in bicycle lanes and pedestrian pathways, make access to employment and social opportunities more affordable and efficient. It also offers a practical opportunity to save people's time and household income as well as government budgets, making investment in sustainable transport a 'win-win' opportunity.

Some western countries are making transportation more sustainable in both long-term and short-term implementations. An example is the modifications in available transportation in Freiburg, Germany. The city has implemented extensive methods of public transportation, cycling, and walking, along with large areas where cars are not allowed.

Since many western countries are highly automobile-orientated areas, the main transit that people use is personal vehicles. About 80% of their travel involves cars. Therefore,

California, is one of the highest greenhouse gases emitters in the United States. The federal government has to come up with some plans to reduce the total number of vehicle trips in order to lower greenhouse gases emission. Such as:

- Improve public transit through the provision of larger coverage area in order to provide more mobility and accessibility, new technology to provide a more reliable and responsive public transportation network.

- Encourage walking and biking through the provision of wider pedestrian pathway, bike share station in commercial downtown, locate parking lot far from the shopping center, limit on street parking, slower traffic lane in downtown area.

- Increase the cost of car ownership and gas taxes through increased parking fees and tolls, encouraging people to drive more fuel efficient vehicles. They can produce social equity problem, since lower people usually drive older vehicles with lower fuel efficiency. Government can use the extra revenue collected from taxes and tolls to improve the public transportation and benefit the poor community.

Other states and nations have built efforts to translate knowledge in behavioral economics into evidence-based sustainable transportation policies.

Business

The most broadly accepted criterion for corporate sustainability constitutes a firm's efficient use of natural capital. This eco-efficiency is usually calculated as the economic value added by a firm in relation to its aggregated ecological impact. This idea has been popularised by the World Business Council for Sustainable Development (WBCSD) under the following definition: "Eco-efficiency is achieved by the delivery of competitively priced goods and services that satisfy human needs and bring quality of life, while progressively reducing ecological impacts and resource intensity throughout the life-cycle to a level at least in line with the earth's carrying capacity." (DeSimone and Popoff, 1997: 47)

Similar to the eco-efficiency concept but so far less explored is the second criterion for corporate sustainability. Socio-efficiency describes the relation between a firm's value added and its social impact. Whereas, it can be assumed that most corporate impacts on the environment are negative (apart from rare exceptions such as the planting of trees) this is not true for social impacts. These can be either positive (e.g. corporate giving, creation of employment) or negative (e.g. work accidents, mobbing of employees, human rights abuses). Depending on the type of impact socio-efficiency thus either tries to minimise negative social impacts (i.e. accidents per value added) or maximise positive social impacts (i.e. donations per value added) in relation to the value added.

Both eco-efficiency and socio-efficiency are concerned primarily with increasing economic sustainability. In this process they instrumentalise both natural and social capital aiming to benefit from win-win situations. However, as Dyllick and Hockerts point out the business case alone will not be sufficient to realise sustainable development. They point towards eco-effectiveness, socio-effectiveness, sufficiency, and eco-equity as four criteria that need to be met if sustainable development is to be reached.

Income

At the present time, sustainable development, along with the solidarity called for in Catholic social teaching, can reduce poverty. While over many thousands of years the 'stronger' (economically or physically) overcame the weaker, nowadays for various reasons - Catholic social teaching, social solidarity, sustainable development – the stronger helps the weaker. This aid may take various forms. 'The Stronger' offers real help rather than striving for the elimination or annihilation of the other. Sustainable development reduces poverty through financial (among other things, a balanced budget), environmental (living conditions), and social (including equality of income) means.

Architecture

In sustainable architecture the recent movements of New Urbanism and New Classical architecture promote a sustainable approach towards construction, that appreciates and develops smart growth, architectural tradition and classical design. This in contrast to modernist and International Style architecture, as well as opposing to solitary housing estates and suburban sprawl, with long commuting distances and large ecological footprints. Both trends started in the 1980s. (It should be noted that sustainable architecture is predominantly relevant to the economics domain while architectural landscaping pertains more to the ecological domain.)

Politics

A study concluded that social indicators and, therefore, sustainable development indicators, are scientific constructs whose principal objective is to inform public policy-making. The International Institute for Sustainable Development has similarly developed a political policy framework, linked to a sustainability index for establishing measurable entities and metrics. The framework consists of six core areas, international trade and investment, economic policy, climate change and energy, measurement and assessment, natural resource management, and the role of communication technologies in sustainable development.

The United Nations Global Compact Cities Programme has defined sustainable political development is a way that broadens the usual definition beyond states and governance. The political is defined as the domain of practices and meanings associated with basic issues of social power as they pertain to the organisation, authorisation, legitimation and regulation of a social life held in common. This definition is in accord with the view that political change is important for responding to economic, ecological and cultural challenges. It also means that the politics of economic change can be addressed. They have listed seven subdomains of the domain of politics:

1. Organization and governance

2. Law and justice

3. Communication and critique

4. Representation and negotiation

5. Security and accord

6. Dialogue and reconciliation

7. Ethics and accountability

This accords with the Brundtland Commission emphasis on development that is guided by human rights principles.

Culture

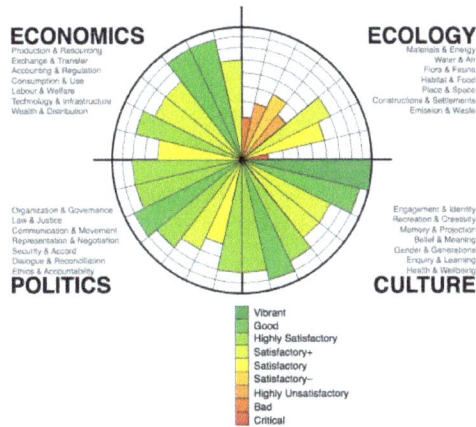

CIRCLES OF SUSTAINABILITY

Framing of sustainable development progress according
to the Circles of Sustainability, used by the United Nations.

Working with a different emphasis, some researchers and institutions have pointed out that a fourth dimension should be added to the dimensions of sustainable development, since the triple-bottom-line dimensions of economic, environmental and social do not seem to be enough to reflect the complexity of contemporary society. In this context, the Agenda 21 for culture and the United Cities and Local Governments (UCLG) Executive Bureau lead the preparation of the policy statement "Culture: Fourth Pillar of Sustainable Development", passed on 17 November 2010, in the framework of the World Summit of Local and Regional Leaders – 3rd World Congress of UCLG, held in Mexico City. although some which still argue that economics is primary, and culture and politics should be included in 'the social'. This document inaugurates a new perspective and points to the relation between culture and sustainable development through a dual approach: developing a solid cultural policy and advocating a cultural dimension in all public policies. The Circles of Sustainability approach distinguishes the four domains of economic, ecological, political and cultural sustainability.

Other organizations have also supported the idea of a fourth domain of sustainable development. The Network of Excellence "Sustainable Development in a Diverse World", sponsored by the European Union, integrates multidisciplinary capacities and interprets cultural diversity as a key element of a new strategy for sustainable development. The Fourth Pillar of Sustainable Development Theory has been referenced by executive director of IMI Institute at UNESCO Vito Di Bari in his manifesto of art and architectural movement Neo-Futurism, whose name was inspired by the 1987 United Nations' report Our Common Future. The Circles of Sustainability approach used by Metropolis defines the (fourth) cultural domain as practices, discourses, and material expressions, which, over time, express continuities and discontinuities of social meaning.

Themes

Progress

The United Nations Conference on Sustainable Development (UNCSD; also known as Rio 2012) was the third international conference on sustainable development, which aimed at reconciling the economic and environmental goals of the global community. An outcome of this conference was the development of the Sustainable Development Goals that aim to promote sustainable progress and eliminate inequalities around the world. However, few nations met the World Wide Fund for Nature's definition of sustainable development criteria established in 2006. Although some nations are more developed than others, all nations are constantly developing because each nation struggles with perpetuating disparities, inequalities and unequal access to fundamental rights and freedoms.

Measurement

In 2007 a report for the U.S. Environmental Protection Agency stated: "While much discussion and effort has gone into sustainability indicators, none of the resulting systems clearly tells us whether our society is sustainable. At best, they can tell us that we are heading in the wrong direction, or that our current activities are not sustainable. More often, they simply draw our attention to the existence of problems, doing little to tell us the origin of those problems and nothing to tell us how to solve them." Nevertheless, a majority of authors assume that a set of well defined and harmonised indicators is the only way to make sustainability tangible. Those indicators are expected to be identified and adjusted through empirical observations (trial and error).

The most common critiques are related to issues like data quality, comparability, objective function and the necessary resources. However a more general criticism is coming from the project management community: How can a sustainable development be achieved at global level if we cannot monitor it in any single project?

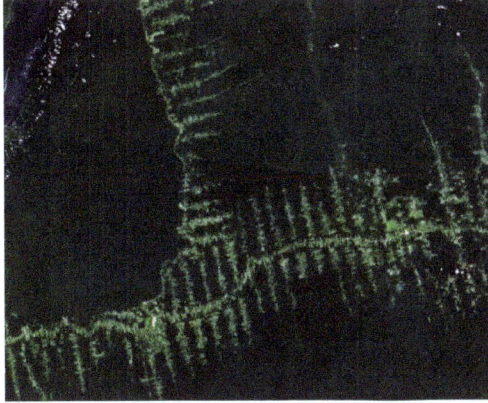

Deforestation and increased road-building in the Amazon Rainforest are a concern because of increased human encroachment upon wilderness areas, increased resource extraction and further threats to biodiversity.

The Cuban-born researcher and entrepreneur Sonia Bueno suggests an alternative approach that is based upon the integral, long-term cost-benefit relationship as a measure and monitoring tool for the sustainability of every project, activity or enterprise. Furthermore, this concept aims to be a practical guideline towards sustainable development following the principle of conservation and increment of value rather than restricting the consumption of resources.

Reasonable qualifications of sustainability are seen U.S. Green Building Council's (USGBC) Leadership in Energy and Environmental Design (LEED). This design incorporates some ecological, economic, and social elements. The goals presented by LEED design goals are sustainable sites, water efficiency, energy and atmospheric emission reduction, material and resources efficiency, and indoor environmental quality. Although amount of structures for sustainability development is many, these qualification has become a standard for sustainable building.

Recent research efforts created also the SDEWES Index to benchmark the performance of cities across aspects that are related to energy, water and environment systems. The SDEWES Index consists of 7 dimensions, 35 indicators, and close to 20 sub-indicators. It is currently applied to 58 cities.

Natural Capital

The sustainable development debate is based on the assumption that societies need to manage three types of capital (economic, social, and natural), which may be non-substitutable and whose consumption might be irreversible. Leading ecological economist and steady-state theorist Herman Daly, for example, points to the fact that natural capital can not necessarily be substituted by economic capital. While it is possible that we can find ways to replace some natural resources, it is much more unlikely that they will ever be able to replace eco-system services, such as the protection provided by the ozone layer, or the climate stabilizing function of the Amazonian forest. In fact natural capital, social capital and economic capital are often complementarities. A further obstacle to substitutability lies also in the multi-functionality of many natural resources. Forests, for example, not only provide the raw material for paper (which can be substituted quite easily), but they also maintain biodiversity, regulate water flow, and absorb CO_2.

Deforestation of native rain forest in Rio de Janeiro City for extraction of clay for civil engineering (2009 picture).

Another problem of natural and social capital deterioration lies in their partial irreversibility. The loss in biodiversity, for example, is often definite. The same can be true for cultural diversity. For example, with globalisation advancing quickly the number of indigenous languages is dropping at alarming rates. Moreover, the depletion of natural and social capital may have non-linear consequences. Consumption of natural and social capital may have no observable impact until a certain threshold is reached. A lake can, for example, absorb nutrients for a long time while actually increasing its productivity. However, once a certain level of algae is reached lack of oxygen causes the lake's ecosystem to break down suddenly.

Business-as-usual

If the degradation of natural and social capital has such important consequence the question arises why action is not taken more systematically to alleviate it. Cohen and Winn point to four types of market failure as possible explanations: First, while the benefits of natural or social capital depletion can usually be privatised, the costs are often externalised (i.e. they are borne not by the party responsible but by society in general). Second, natural capital is often undervalued by society since we are not fully aware of the real cost of the depletion of natural capital. Information asymmetry is a third reason—often the link between cause and effect is obscured, making it difficult for actors to make informed choices. Cohen and Winn close with the realization that contrary to economic theory many firms are not perfect optimisers. They postulate that firms often do not optimise resource allocation because they are caught in a "business as usual" mentality.

Before flue-gas desulfurization was installed, the air-polluting emissions from this power plant in New Mexico contained excessive amounts of sulfur dioxide.

Criticism

It has been argued that since the 1960s, the concept of sustainable development has changed from 'conservation management' to 'economic development', whereby the original meaning of the concept has been stretched somewhat.

In the 1960s, the international community realised that many African countries needed national plans to safeguard wildlife habitats, and that rural areas had to confront the limits imposed by soil, climate and water availability. This was a strategy of conservation management. In the 70s, however, the focus shifted to the broader issues of the provisioning of basic human needs, community participation as well as appropriate technology use throughout the developing countries (and not just in Africa). This was a strategy of economic development, and the strategy was carried even further by the Brundtland Report when the issues went from regional to international in scope and application. In effect, the conservationists were crowded out and superseded by the developers.

But shifting the focus of sustainable development from conservation to development has had the imperceptible effect of stretching the original forest management term of sustainable yield from the use of renewable resources only (like forestry), to now also accounting for the use of non-renewable resources (like minerals). This stretching of the term has been questioned. Thus, environmental economist Kerry Turner has argued that literally, there can be no such thing as overall 'sustainable development' in an industrialised world economy that remains heavily dependent on the extraction of Earth's finite stock of exhaustible mineral resources:

> " It makes no sense to talk about the sustainable use of a non-renewable resource (even with substantial recycling effort and use rates). Any positive rate of exploitation will eventually lead to exhaustion of the finite stock. "

In effect, it has been argued that the Industrial Revolution as a whole is unsustainable.

Sustainability

In ecology, sustainability is the combination of the words sustain and ability. This concept makes reference to how biological systems remain diverse and productive indefinitely. Long-lived and healthy wetlands and forests are examples of sustainable biological systems. In more general terms, sustainability is the endurance of systems and processes. The organizing principle for sustainability is sustainable development, which includes the four interconnected domains: ecology, economics, politics and culture. Sustainability science is the study of sustainable development and environmental science.

Achieving sustainability will enable the Earth to continue supporting human life.

Sustainability can also be defined as a socio-ecological process characterized by the pursuit of a common ideal. An ideal is by definition unattainable in a given time/space but endlessly approachable and it is this endless pursuit what builds in sustainability in the process (ibid). Healthy ecosystems and environments are necessary to the survival of humans and other organisms. Ways of reducing negative human impact are environmentally-friendly chemical engineering, environmental resources management and environmental protection. Information is gained from green chemistry, earth science, environmental science and conservation biology. Ecological economics studies the fields of academic research that aim to address human economies and natural ecosystems.

Moving towards sustainability is also a social challenge that entails international and national law, urban planning and transport, local and individual lifestyles and ethical consumerism. Ways of living more sustainably can take many forms from reorganizing living conditions (e.g., ecovillages, eco-municipalities and sustainable cities), reappraising economic sectors (permaculture, green building, sustainable agriculture), or work practices (sustainable architecture), using science to develop new technologies (green technologies, renewable energy and sustainable fission and fusion power), or designing systems in a flexible and reversible manner, and adjusting individual lifestyles that conserve natural resources.

Batad rice terraces, The Philippines —UNESCO World Heritage site

Despite the increased popularity of the use of the term "sustainability", the possibility that human societies will achieve environmental sustainability has been, and continues to be, questioned—in light of environmental degradation, climate change, overconsumption, population growth and societies' pursuit of indefinite economic growth in a closed system.

Etymology

The name sustainability is derived from the Latin *sustinere* (*tenere*, to hold; *sub*, up). *Sustain* can mean "maintain", "support", or "endure". Since the 1980s *sustainability* has been used more in the sense of human sustainability on planet Earth and this has resulted in the most widely quoted definition of sustainability as a part of the concept *sustainable development*, that of the Brundtland Commission of the United Nations on March 20, 1987: "sustainable development is development that meets the needs of the present without compromising the ability of future generations to meet their own needs."

Components

Three Pillars of Sustainability

The 2005 World Summit on Social Development identified sustainable development goals, such as economic development, social development and environmental protection. This view has been expressed as an illustration using three overlapping ellipses indicating that the three pillars of sustainability are not mutually exclusive and can be mutually reinforcing. In fact, the three pillars are interdependent, and in the long run none can exist without the others. The three pillars have served as a common ground for numerous sustainability standards and certification systems in recent years, in particular in the food industry. Standards which today explicitly refer to the triple bottom line include Rainforest Alliance, Fairtrade and UTZ Certified. Some sustainability experts and practitioners have illustrated four pillars of sustainability, or a quadruple bottom line. One such pillar is future generations, which emphasizes the long-term thinking associated with sustainability.

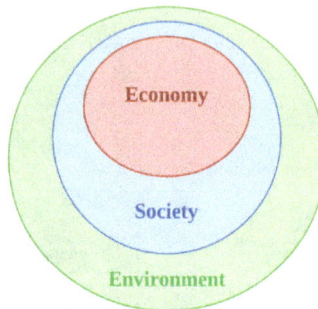

A diagram indicating the relationship between the "three pillars of sustainability", in which both economy and society are constrained by environmental limits

Sustainable development consists of balancing local and global efforts to meet basic human needs without destroying or degrading the natural environment. The question then becomes how to represent the relationship between those needs and the environment.

A study from 2005 pointed out that environmental justice is as important as is sustainable development. Ecological economist Herman Daly asked, "what use is a sawmill without a forest?" From this perspective, the economy is a subsystem of human society, which is itself a subsystem of the biosphere, and a gain in one sector is a loss from another. This perspective led to the nested circles figure of 'economics' inside 'society' inside the 'environment'.

The simple definition that sustainability is something that improves "the quality of human life while living within the carrying capacity of supporting eco-systems", though vague, conveys the idea of sustainability having quantifiable limits. But sustainability is also a call to action, a task in progress or "journey" and therefore a political process, so some definitions set out common goals and values. The Earth Charter speaks of "a sustainable global society founded on respect for nature, universal human rights, economic justice, and a culture of peace." This suggested a more complex figure of sustainability, which included the importance of the domain of 'politics'.

More than that, sustainability implies responsible and proactive decision-making and innovation that minimizes negative impact and maintains balance between ecological resilience, economic prosperity, political justice and cultural vibrancy to ensure a desirable planet for all species now and in the future. Specific types of sustainability include, sustainable agriculture, sustainable architecture or ecological economics. Understanding sustainable development is important but without clear targets an unfocused term like "liberty" or "justice". It has also been described as a "dialogue of values that challenge the sociology of development".

Circles of Sustainability

While the United Nations Millennium Declaration identified principles and treaties on sustainable development, including economic development, social development and environmental protection it continued using three domains: economics, environment and social sustainability. More recently, using a systematic domain model that responds to the debates over the last decade, the Circles of Sustainability approach distinguished four domains of economic, ecological, political and cultural sustainability. This in accord with the United Nations Agenda 21, which specifies culture as the fourth domain of sustainable development. The model is now being used by organizations such as the United Nations Cities Programme. and Metropolis

Urban sustainability analysis of the greater urban area of the city of São Paulo using the 'Circles of Sustainability' method of the UN and Metropolis Association.

Shaping The Future

Integral elements of sustainability are research and innovation activities. A telling example is the European environmental research and innovation policy. It aims at defining and implementing a transformative agenda to greening the economy and the society as a whole so to make them sustainable. Research and innovation in Europe are financially supported by the programme Horizon 2020, which is also open to participation worldwide.

Resiliency

Resiliency in ecology is the capacity of an ecosystem to absorb disturbance and still retain its basic structure and viability. Resilience-thinking evolved from the need to manage interactions between human-constructed systems and natural ecosystems in a sustainable way despite the fact that to policymakers a definition remains elusive. Resilience-thinking addresses how much planetary ecological systems can withstand assault from human disturbances and still deliver the services current and future generations need from them. It is also concerned with commitment from geopolitical policymakers to promote and manage essential planetary ecological resources in order to promote resilience and achieve sustainability of these essential resources for benefit of future generations of life? The resiliency of an ecosystem, and thereby, its sustainability, can be reasonably measured at junctures or events where the combination of naturally occurring regenerative forces (solar energy, water, soil, atmosphere, vegetation, and biomass) interact with the energy released into the ecosystem from disturbances.

A practical view of sustainability is closed systems that maintain processes of productivity indefinitely by replacing resources used by actions of people with resources of equal or greater value by those same people without degrading or endangering natural biotic systems. In this way, sustainability can be concretely measured in human projects if there is a transparent accounting of the resources put back into the ecosystem to replace those displaced. In nature, the accounting occurs naturally through a process of adaptation as an ecosystem returns to viability from an external disturbance. The adaptation is a multi-stage process that begins with the disturbance event (earthquake, volcanic eruption, hurricane, tornado, flood, or thunderstorm), followed by absorption, utilization, or deflection of the energy or energies that the external forces created.

In analysing systems such as urban and national parks, dams, farms and gardens, theme parks, open-pit mines, water catchments, one way to look at the relationship between sustainability and resiliency is to view the former with a long-term vision and resiliency as the capacity of human engineers to respond to immediate environmental events.

History

The history of sustainability traces human-dominated ecological systems from the earliest civilizations to the present time. This history is characterized by the increased regional success of a particular society, followed by crises that were either resolved, producing sustainability, or not, leading to decline.

In early human history, the use of fire and desire for specific foods may have altered the natural composition of plant and animal communities. Between 8,000 and 10,000 years ago, agrarian communities emerged which depended largely on their environment and the creation of a "structure of permanence."

The Western industrial revolution of the 18th to 19th centuries tapped into the vast growth potential of the energy in fossil fuels. Coal was used to power ever more efficient engines and later to generate electricity. Modern sanitation systems and advances in medicine protected large populations from disease. In the mid-20th century, a gathering environmental movement pointed out that there were environmental costs associated with the many material benefits that were now being enjoyed. In the late 20th century, environmental problems became global in scale. The 1973 and 1979 energy crises demonstrated the extent to which the global community had become dependent on non-renewable energy resources.

In the 21st century, there is increasing global awareness of the threat posed by the human greenhouse effect, produced largely by forest clearing and the burning of fossil fuels.

Principles and Concepts

The philosophical and analytic framework of sustainability draws on and connects with many different disciplines and fields; in recent years an area that has come to be called sustainability science has emerged.

The United Nations Millennium Declaration identified principles and treaties on sustainable development, including economic development, social development and environmental protection. The Circles of Sustainability approach distinguishes the four domains of economic, ecological, political and cultural sustainability. This in accord with the United Nations Agenda 21, which specifies culture as the fourth domain of sustainable development.

Scale and Context

Sustainability is studied and managed over many scales (levels or frames of reference) of time and space and in many contexts of environmental, social and economic organization. The focus ranges from the total carrying capacity (sustainability) of planet Earth to the sustainability of economic sectors, ecosystems, countries, municipalities, neighbourhoods, home gardens, individual lives, individual goods and services, occupations, lifestyles, behaviour patterns and so on. In short, it can entail the full compass of biological and human activity or any part of it. As Daniel Botkin, author and environmentalist, has stated: "We see a landscape that is always in flux, changing over many scales of time and space."

The sheer size and complexity of the planetary ecosystem has proved problematic for the design of practical measures to reach global sustainability. To shed light on the big picture, explorer and sustainability campaigner Jason Lewis has drawn parallels to other, more tangible closed systems. For example, he likens human existence on Earth — isolated as the planet is in space, whereby people cannot be evacuated to relieve population pressure and resources cannot be imported to prevent accelerated depletion of resources — to life at sea on a small boat isolated by water. In both cases, he argues, exercising the precautionary principle is a key factor in survival.

Consumption

A major driver of human impact on Earth systems is the destruction of biophysical resources, and especially, the Earth's ecosystems. The environmental impact of a community or of humankind as a whole depends both on population and impact per person, which in turn depends in complex ways on what resources are being used, whether or not those resources are renewable, and the scale of the human activity relative to the carrying capacity of the ecosystems involved. Careful resource management can be applied at many scales, from economic sectors like agriculture, manufacturing and industry, to work organizations, the consumption patterns of households and individuals and to the resource demands of individual goods and services.

One of the initial attempts to express human impact mathematically was developed in the 1970s and is called the I PAT formula. This formulation attempts to explain human consumption in terms of three components: population numbers, levels of consumption (which it terms "affluence", although the usage is different), and impact per unit of resource use (which is termed "technology", because this impact depends on the technology used). The equation is expressed:

$$I = P \times A \times T$$

Where: I = Environmental impact, P = Population, A = Affluence, T = Technology

Measurement

Sustainability measurement is a term that denotes the measurements used as the quantitative basis for the informed management of sustainability. The metrics used for the measurement of sustainability (involving the sustainability of environmental, social and economic domains, both individually and in various combinations) are evolving: they include indicators, benchmarks, audits, sustainability standards and certification systems like Fairtrade and Organic, indexes and accounting, as well as assessment, appraisal and other reporting systems. They are applied over a wide range of spatial and temporal scales.

Some of the best known and most widely used sustainability measures include corporate sustainability reporting, Triple Bottom Line accounting, World Sustainability Society, Circles of Sustainability, and estimates of the quality of sustainability governance for individual countries using the Environmental Sustainability Index and Environmental Performance Index.

Population

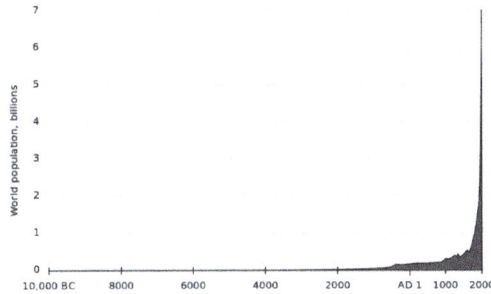

Graph showing human population growth from 10,000 BC – 2000 AD, illustrating current exponential growth

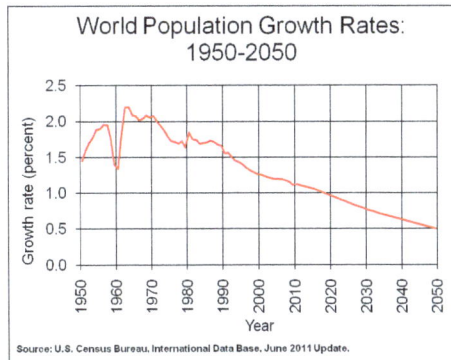

World population growth rate, 1950–2050, as estimated in 2011 by the U.S. Census Bureau, International Data Base

According to the 2008 Revision of the official United Nations population estimates and projections, the world population is projected to reach 7 billion early in 2012, up from the current 6.9 billion (May 2009), to exceed 9 billion people by 2050. Most of the increase will be in developing countries whose population is projected to rise from 5.6 billion in 2009 to 7.9 billion in 2050. This increase will be distributed among the population aged 15–59 (1.2 billion) and 60 or over (1.1 billion) because the number of children under age 15 in developing countries is predicted to decrease. In contrast, the population of the more developed regions is expected to undergo only slight increase from 1.23 billion to 1.28 billion, and this would have declined to 1.15 billion but for a projected net migration from developing to developed countries, which is expected to average 2.4 million persons annually from 2009 to 2050. Long-term estimates in 2004 of global population suggest a peak at around 2070 of nine to ten billion people, and then a slow decrease to 8.4 billion by 2100.

Emerging economies like those of China and India aspire to the living standards of the Western world as does the non-industrialized world in general. It is the combination of population increase in the developing world and unsustainable consumption levels in the developed world that poses a stark challenge to sustainability.

Carrying Capacity

At the global scale, scientific data now indicates that humans are living beyond the carrying capacity of planet Earth and that this cannot continue indefinitely. This scientific evidence comes from many sources but is presented in detail in the Millennium Ecosystem Assessment and the planetary boundaries framework. An early detailed examination of global limits was published in the 1972 book *Limits to Growth*, which has prompted follow-up commentary and analysis. A 2012 review in *Nature* by 22 international researchers expressed concerns that the Earth may be "approaching a state shift" in its biosphere.

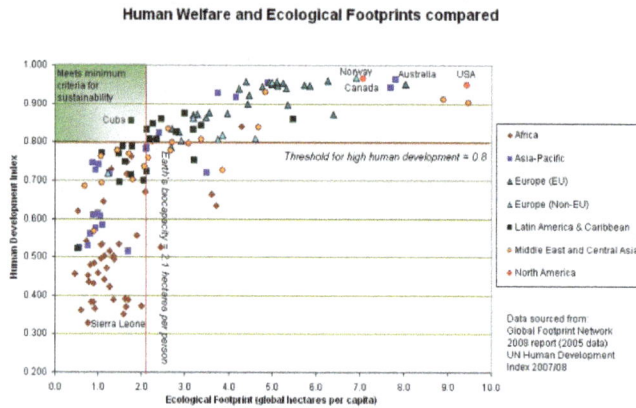

Ecological footprint for different nations compared to their Human Development Index (HDI)

The Ecological footprint measures human consumption in terms of the biologically productive land needed to provide the resources, and absorb the wastes of the average global citizen. In 2008 it required 2.7 global hectares per person, 30% more than the natural biological capacity of 2.1 global hectares (assuming no provision for other organisms). The resulting ecological deficit must be met from unsustainable *extra* sources and these are obtained in three ways: embedded in the goods and services of world trade; taken from the past (e.g. fossil fuels); or borrowed from the future as unsustainable resource usage (e.g. by over exploiting forests and fisheries).

The figure (right) examines sustainability at the scale of individual countries by contrasting their Ecological Footprint with their UN Human Development Index (a measure of standard of living). The graph shows what is necessary for countries to maintain an acceptable standard of living for their citizens while, at the same time, maintaining sustainable resource use. The general trend is for higher standards of living to become less sustainable. As always, population growth has a marked influence on levels of consumption and the efficiency of resource use. The sustainability goal is to raise the global standard of living without increasing the use of resources beyond globally sustainable levels; that is, to not exceed "one planet" consumption. Information generated by reports at the national, regional and city scales confirm the global trend towards societies that are becoming less sustainable over time.

Romanian American economist Nicholas Georgescu-Roegen, a progenitor in economics and a paradigm founder of ecological economics, has argued that the carrying capacity of Earth — that is, Earth's capacity to sustain human populations and consumption levels — is bound to decrease sometime in the future as Earth's finite stock of mineral resources is presently being extracted and put to use. Leading ecological economist and steady-state theorist Herman Daly, a student of Georgescu-Roegen, has propounded the same argument.

Global Human Impact on Biodiversity

At a fundamental level energy flow and biogeochemical cycling set an upper limit on the number and mass of organisms in any ecosystem. Human impacts on the Earth are demonstrated in a general way through detrimental changes in the global biogeochemical cycles of chemicals that are critical to life, most notably those of water, oxygen, carbon, nitrogen and phosphorus.

The *Millennium Ecosystem Assessment* is an international synthesis by over 1000 of the world's leading biological scientists that analyzes the state of the Earth's ecosystems and provides summaries and guidelines for decision-makers. It concludes that human activity is having a significant and escalating impact on the biodiversity of world ecosystems, reducing both their resilience and biocapacity. The report refers to natural systems as humanity's "life-support system", providing essential "ecosystem services". The assessment measures 24 ecosystem services concluding that only four have shown improvement over the last 50 years, 15 are in serious decline, and five are in a precarious condition.

Sustainable Development Goals

The Sustainable Development Goals (SDGs) are the current harmonized set of seventeen future international development targets.

The Official Agenda for Sustainable Development adopted on 25 September 2015 has 92 paragraphs, with the main paragraph (51) outlining the 17 Sustainable Development Goals and its associated 169 targets. This included the following seventeen goals:

1. Poverty – End poverty in all its forms everywhere

2. Food – End hunger, achieve food security and improved nutrition and promote sustainable agriculture

3. Health – Ensure healthy lives and promote well-being for all at all ages

4. Education – Ensure inclusive and equitable quality education and promote life-long learning opportunities for all

5. Women – Achieve gender equality and empower all women and girls

6. Water – Ensure availability and sustainable management of water and sanitation for all

7. Energy – Ensure access to affordable, reliable, sustainable and modern energy for all

8. Economy – Promote sustained, inclusive and sustainable economic growth, full and productive employment and decent work for all

9. Infrastructure – Build resilient infrastructure, promote inclusive and sustainable industrialization and foster innovation

10. Inequality – Reduce inequality within and among countries

11. Habitation – Make cities and human settlements inclusive, safe, resilient and sustainable

12. Consumption – Ensure sustainable consumption and production patterns

13. Climate – Take urgent action to combat climate change and its impacts

14. Marine-ecosystems – Conserve and sustainably use the oceans, seas and marine resources for sustainable development

15. Ecosystems – Protect, restore and promote sustainable use of terrestrial ecosystems, sustainably manage forests, combat desertification, and halt and reverse land degradation and halt biodiversity loss

16. Institutions – Promote peaceful and inclusive societies for sustainable development, provide access to justice for all and build effective, accountable and inclusive institutions at all levels

17. Sustainability – Strengthen the means of implementation and revitalize the global partnership for sustainable development

As of August 2015, there were 169 proposed targets for these goals and 304 proposed indicators to show compliance.

The Sustainable Development Goals (SDGs) replace the eight Millennium Development Goals (MDGs), which expired at the end of 2015. The MDGs were established in 2000 following the Millennium Summit of the United Nations. Adopted by the 189 United Nations member states at the time and more than twenty international organizations, these goals were advanced to help achieve the following sustainable development standards by 2015.

1. To eradicate extreme poverty and hunger

2. To achieve universal primary education

3. To promote gender equality and empower women

4. To reduce child mortality

5. To improve maternal health

6. To combat HIV/AIDS, malaria, and other diseases

7. To ensure environmental sustainability (one of the targets in this goal focuses on increasing sustainable access to safe drinking water and basic sanitation)

8. To develop a global partnership for development

Sustainable Development

According to the data that member countries represented to the United Nations, Cuba was the only nation in the world in 2006 that met the World Wide Fund for Nature's definition of sustainable development, with an ecological footprint of less than 1.8 hectares per capita, 1.5, and a Human Development Index of over 0.8, 0.855.

Environmental Dimension

Healthy ecosystems provide vital goods and services to humans and other organisms. There are two major ways of reducing negative human impact and enhancing ecosystem services and the first of these is environmental management. This direct approach is based largely on information gained from earth science, environmental science and conservation biology. However, this is management at the end of a long series of indirect causal factors that are initiated by human consumption, so a second approach is through demand management of human resource use.

Management of human consumption of resources is an indirect approach based largely on information gained from economics. Herman Daly has suggested three broad criteria for ecological sustainability: renewable resources should provide a sustainable yield (the rate of harvest should not exceed the rate of regeneration); for non-renewable resources there should be equivalent development of renewable substitutes; waste generation should not exceed the assimilative capacity of the environment.

Environmental Management

At the global scale and in the broadest sense environmental management involves the oceans, freshwater systems, land and atmosphere, but following the sustainability principle of scale it can be equally applied to any ecosystem from a tropical rainforest to a home garden.

Atmosphere

At a March 2009 meeting of the Copenhagen Climate Council, 2,500 climate experts from 80 countries issued a keynote statement that there is now "no excuse" for failing to act on global warming and that without strong carbon reduction "abrupt or irreversible" shifts in climate may occur that "will be very difficult for contemporary societies to cope with". Management of the global atmosphere now involves assessment of all aspects of the carbon cycle to identify opportunities to address human-induced climate change and this has become a major focus of scientific research because of the potential catastrophic effects on biodiversity and human communities.

Other human impacts on the atmosphere include the air pollution in cities, the pollutants including toxic chemicals like nitrogen oxides, sulfur oxides, volatile organic compounds and airborne particulate matter that produce photochemical smog and acid rain, and the chlorofluorocarbons that degrade the ozone layer. Anthropogenic particulates such as sulfate aerosols in the atmosphere reduce the direct irradiance and reflectance (albedo) of the Earth's surface. Known as global dimming, the decrease is estimated to have been about 4% between 1960 and 1990 although the trend has subsequently reversed. Global dimming may have disturbed the global water cycle by reducing evaporation and rainfall in some areas. It also creates a cooling effect and this may have partially masked the effect of greenhouse gases on global warming.

Freshwater and Oceans

Water covers 71% of the Earth's surface. Of this, 97.5% is the salty water of the oceans and only 2.5% freshwater, most of which is locked up in the Antarctic ice sheet. The remaining freshwater is found in glaciers, lakes, rivers, wetlands, the soil, aquifers and atmosphere. Due to the water cycle, fresh water supply is continually replenished by precipitation, however there is still a limited amount necessitating management of this resource. Awareness of the global importance of preserving water for ecosystem services has only recently emerged as, during the 20th century, more than half the world's wetlands have been lost along with their valuable environmental services. Increasing urbanization pollutes clean water supplies and much of the world still does not have access to clean, safe water. Greater emphasis is now being placed on the improved management of blue (harvestable) and green (soil water available for plant use) water, and this applies at all scales of water management.

Ocean circulation patterns have a strong influence on climate and weather and, in turn, the food supply of both humans and other organisms. Scientists have warned of the possibility, under the influence of climate change, of a sudden alteration in circulation patterns of ocean currents that could drastically alter the climate in some regions of the globe. Ten per cent of the world's population – about 600 million people – live in low-lying areas vulnerable to sea level rise.

Land Use

Loss of biodiversity stems largely from the habitat loss and fragmentation produced by the human appropriation of land for development, forestry and agriculture as natural capital is progressively converted to man-made capital. Land use change is fundamental to the operations of the biosphere because alterations in the relative proportions of land dedicated to urbanisation, agriculture, forest, woodland, grassland and pasture have a marked effect on the global water, carbon and nitrogen biogeochemical cycles and this can impact negatively on both natural and human systems. At the local human scale, major sustainability benefits accrue from sustainable parks and gardens and green cities.

A rice paddy in Bangladesh. Rice, wheat, corn and potatoes make up more than half the world's food supply.

Since the Neolithic Revolution about 47% of the world's forests have been lost to human use. Present-day forests occupy about a quarter of the world's ice-free land with about half of these occurring in the tropics. In temperate and boreal regions forest area is gradually increasing (with the exception of Siberia), but deforestation in the tropics is of major concern.

Food is essential to life. Feeding more than seven billion human bodies takes a heavy toll on the Earth's resources. This begins with the appropriation of about 38% of the Earth's land surface and about 20% of its net primary productivity. Added to this are the resource-hungry activities of industrial agribusiness – everything from the crop need for irrigation water, synthetic fertilizers and pesticides to the resource costs of food packaging, transport (now a major part of global trade) and retail. Environmental problems associated with industrial agriculture and agribusiness are now being addressed through such movements as sustainable agriculture, organic farming and more sustainable business practices.

Management of Human Consumption

The underlying driver of direct human impacts on the environment is human consumption. This impact is reduced by not only consuming less but by also making the full cycle of production, use and disposal more sustainable. Consumption of goods and services can be analysed and managed at all scales through the chain of consumption, starting with the effects of individual lifestyle choices and spending patterns, through to the resource demands of specific goods and services, the impacts of economic sectors, through national economies to the global economy. Analysis of consumption patterns relates resource use to the environmental, social and economic impacts at the scale or context under investigation. The ideas of embodied resource use (the total resources needed to produce a product or service), resource intensity, and resource productivity are important tools for understanding the impacts of consumption. Key resource categories relating to human needs are food, energy, materials and water.

The Helix of Sustainability

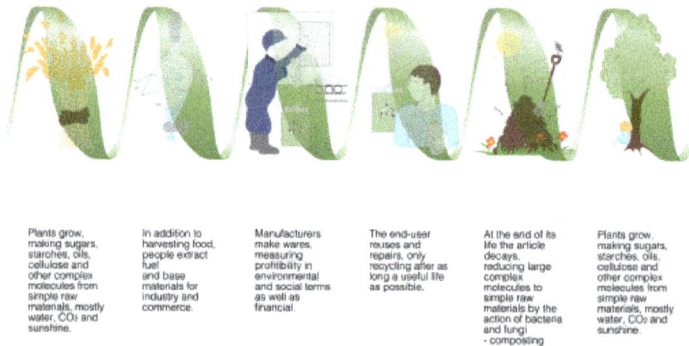

| Plants grow, making sugars, starches, oils, cellulose and other complex molecules from simple raw materials, mostly water, CO₂ and sunshine. | In addition to harvesting food, people extract fuel and base materials for industry and commerce. | Manufacturers make wares, measuring profitability in environmental and social terms as well as financial. | The end-user reuses and repairs, only recycling after as long a useful life as possible. | At the end of its life the article decays, reducing large complex molecules to simple raw materials by the action of bacteria and fungi - composting | Plants grow, making sugars, starches, oils, cellulose and other complex molecules from simple raw materials, mostly water, CO₂ and sunshine. |

Helix of sustainability – the carbon cycle of manufacturing

In 2010, the International Resource Panel, hosted by the United Nations Environment Programme (UNEP), published the first global scientific assessment on the impacts of consumption and production and identified priority actions for developed and developing countries. The study found that the most critical impacts are related to ecosystem health, human health and resource depletion. From a production perspective, it found that fossil-fuel combustion processes, agriculture and fisheries have the most important impacts. Meanwhile, from a final consumption perspective, it found that household consumption related to mobility, shelter, food and energy-using products cause the majority of life-cycle impacts of consumption.

Energy

The Sun's energy, stored by plants (primary producers) during photosynthesis, passes through the food chain to other organisms to ultimately power all living processes. Since the industrial revolution the concentrated energy of the Sun stored in fossilized plants as fossil fuels has been a major driver of technology which, in turn, has been the source of both economic and political power. In 2007 climate scientists of the IPCC concluded that there was at least a 90% probability that atmospheric increase in CO_2 was human-induced, mostly as a result of fossil fuel emissions but, to a lesser extent from changes in land use. Stabilizing the world's climate will require high-income countries to reduce their emissions by 60–90% over 2006 levels by 2050 which should hold CO_2 levels at 450–650 ppm from current levels of about 380 ppm. Above this level, temperatures could rise by more than 2 °C to produce "catastrophic" climate change. Reduction of current CO_2 levels must be achieved against a background of global population increase and developing countries aspiring to energy-intensive high consumption Western lifestyles.

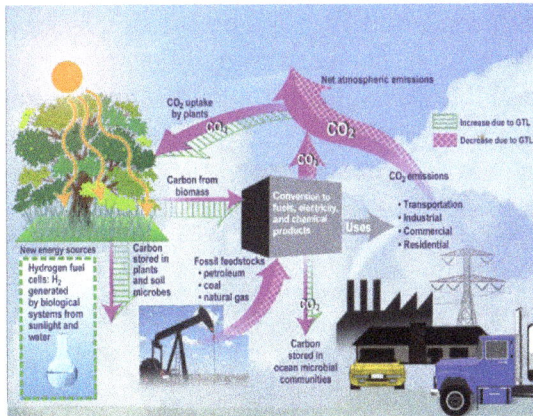

Flow of CO_2 in an ecosystem

Reducing greenhouse emissions, is being tackled at all scales, ranging from tracking the passage of carbon through the carbon cycle to the commercialization of renewable energy, developing less carbon-hungry technology and transport systems and attempts by individuals to lead carbon neutral lifestyles by monitoring the fossil fuel use embodied in all the goods and services they use. Engineering of emerging technologies such as carbon-neutral fuel and energy storage systems such as power to gas, compressed air energy storage, and pumped-storage hydroelectricity are necessary to store power from transient renewable energy sources including emerging renewables such as airborne wind turbines.

Water

Water security and food security are inextricably linked. In the decade 1951–60 human water withdrawals were four times greater than the previous decade. This rapid increase resulted from scientific and technological developments impacting through the economy – especially the increase in irrigated land, growth in industrial and power sectors, and intensive dam construction on all continents. This altered the water cycle of rivers and lakes, affected their water quality and had a significant impact on the global water cycle. Currently towards 35% of human water use is unsustainable, drawing on diminishing aquifers and reducing the flows of major rivers: this percentage is likely to increase if climate change impacts become more severe, populations increase, aquifers become progressively depleted and supplies become polluted and unsanitary. From 1961 to 2001 water demand doubled — agricultural use increased by 75%, industrial use by more than 200%, and domestic use more than 400%. In the 1990s it was estimated that humans were using 40–50% of the globally available freshwater in the approximate proportion of 70% for agriculture, 22% for industry, and 8% for domestic purposes with total use progressively increasing.

Water efficiency is being improved on a global scale by increased demand management, improved infrastructure, improved water productivity of agriculture, minimising the water intensity (embodied water) of goods and services, addressing shortages in the non-industrialized world, concentrating food production in areas of high productivity, and planning for climate change, such as through flexible system design. A promising direction towards sustainable development is to design systems that are flexible and reversible. At the local level, people are becoming more self-sufficient by harvesting rainwater and reducing use of mains water.

Food

Feijoada — A typical black bean food dish from Brazil

The American Public Health Association (APHA) defines a "sustainable food system" as "one that provides healthy food to meet current food needs while maintaining healthy ecosystems that can also provide food for generations to come with minimal negative impact to the environment. A sustainable food system also encourages local production and distribution infrastructures and makes nutritious food available, accessible, and affordable to all. Further, it is humane and just, protecting farmers and other workers, consumers, and communities." Concerns about the environmental impacts of agribusiness and the stark contrast between the obesity problems of the Western world and the poverty and food insecurity of the developing world have generated a strong movement towards healthy, sustainable eating as a major component of overall ethical consumerism. The environmental effects of different dietary patterns depend on many factors, including the proportion of animal and plant foods consumed and the method of food production. The World Health Organization has published a *Global Strategy on Diet, Physical Activity and Health* report which was endorsed by the May 2004 World Health Assembly. It recommends the Mediterranean diet which is associated with health and longevity and is low in meat, rich in fruits and vegetables, low in added sugar and limited salt, and low in saturated fatty acids; the traditional source of fat in the Mediterranean is olive oil, rich in monounsaturated fat. The healthy rice-based Japanese diet is also high in carbohydrates and low in fat. Both diets are low in meat and saturated fats and high in legumes and other vegetables; they are associated with a low incidence of ailments and low environmental impact.

At the global level the environmental impact of agribusiness is being addressed through sustainable agriculture and organic farming. At the local level there are various movements working towards local food production, more productive use of urban wastelands and domestic gardens including permaculture, urban horticulture, local food, slow food, sustainable gardening, and organic gardening.

Sustainable seafood is seafood from either fished or farmed sources that can maintain or increase production in the future without jeopardizing the ecosystems from which it was acquired. The sustainable seafood movement has gained momentum as more people become aware about both overfishing and environmentally destructive fishing methods.

Materials, Toxic Substances, Waste

An electric wire reel reused as a center table in a Rio de Janeiro decoration fair. The reuse of materials is a sustainable practice that is rapidly growing among designers in Brazil.

As global population and affluence has increased, so has the use of various materials increased in volume, diversity and distance transported. Included here are raw materials, minerals, synthetic chemicals (including hazardous substances), manufactured products, food, living organisms and waste. By 2050, humanity could consume an estimated 140 billion tons of minerals, ores, fossil fuels and biomass per year (three times its current amount) unless the economic growth rate is decoupled from the rate of natural resource consumption. Developed countries' citizens consume an average of 16 tons of those four key resources per capita, ranging up to 40 or more tons per person in some developed countries with resource consumption levels far beyond what is likely sustainable.

Sustainable use of materials has targeted the idea of dematerialization, converting the linear path of materials (extraction, use, disposal in landfill) to a circular material flow that reuses materials as much as possible, much like the cycling and reuse of waste in nature. This approach is supported by product stewardship and the increasing use of material flow analysis at all levels, especially individual countries and the global economy. The use of sustainable biomaterials that come from renewable sources and that can be recycled is preferred to the use on non-renewables from a life cycle standpoint.

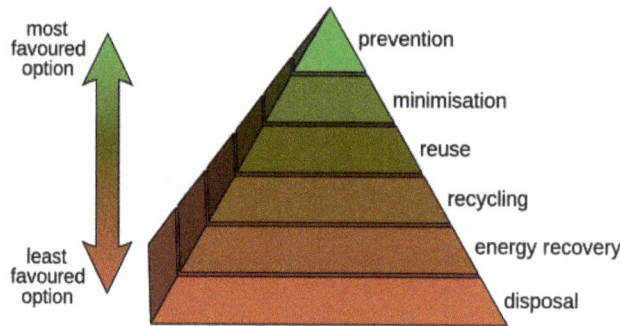

The waste hierarchy

Synthetic chemical production has escalated following the stimulus it received during the second World War. Chemical production includes everything from herbicides, pesticides and fertilizers to domestic chemicals and hazardous substances. Apart from the build-up of greenhouse gas emissions in the atmosphere, chemicals of particular concern include: heavy metals, nuclear waste, chlorofluorocarbons, persistent organic pollutants and all harmful chemicals capable of bioaccumulation. Although most synthetic chemicals are harmless there needs to be rigorous testing of new chemicals, in all countries, for adverse environmental and health effects. International legislation has been established to deal with the global distribution and management of dangerous goods. The effects of some chemical agents needed long-term measurements and a lot of legal battles to realize their danger to human health. The classification of the toxic carcinogenic agents is handle by the International Agency for Research on Cancer.

Every economic activity produces material that can be classified as waste. To reduce waste industry, business and government are now mimicking nature by turning the waste produced by industrial metabolism into resource. Dematerialization is being encouraged through the ideas of industrial ecology, ecodesign and ecolabelling. In addition to the well-established "reduce, reuse and recycle," shoppers are using their purchasing power for ethical consumerism.

The European Union is expected to table by the end of 2015 an ambitious Circular Economy package which is expected to include concrete legislative proposals on waste management, ecodesign and limits on land fills.

Economic Dimension

On one account, sustainability "concerns the specification of a set of actions to be taken by present persons that will not diminish the prospects of future persons to enjoy levels of consumption, wealth, utility, or welfare comparable to those enjoyed by present persons." Sustainability interfaces with economics through the social and ecological consequences of economic activity. Sustainability economics represents: "... a broad interpretation of ecological economics where environmental and ecological variables and issues are basic but part of a multidimensional perspective. Social, cultural, health-related and monetary/financial aspects have to be integrated into the analysis." However, the concept of sustainability is much broader than the concepts of sustained yield of welfare, resources, or profit margins. At present, the average per capita consumption of people in the developing world is sustainable but population numbers are increasing and individuals are aspiring to high-consumption Western lifestyles. The developed world population is only increasing slightly but consumption levels are unsustainable. The challenge for sustainability is to curb and manage Western consumption while raising the standard of living of the developing world without increasing its resource use and environmental impact. This must be done by using strategies and technology that break the link between, on the one hand, economic growth and on the other, environmental damage and resource depletion.

The *Great Fish Market*, painted by Jan Brueghel the Elder

A recent UNEP report proposes a green economy defined as one that "improves human well-being and social equity, while significantly reducing environmental risks and ecological scarcities": it "does not favor one political perspective over another but works to minimize excessive depletion of natural capital". The report makes three key findings: "that greening not only generates increases in wealth, in particular a gain in ecological commons or natural capital, but also (over a period of six years) produces a higher rate of GDP growth"; that there is "an inextricable link between poverty eradication and better maintenance and conservation of the ecological commons, arising from the benefit flows from natural capital that are received directly by the poor"; "in the transition to a green economy, new jobs are created, which in time exceed the losses in "brown economy" jobs. However, there is a period of job losses in transition, which requires investment in re-skilling and re-educating the workforce".

Several key areas have been targeted for economic analysis and reform: the environmental effects of unconstrained economic growth; the consequences of nature being treated as an economic externality; and the possibility of an economics that takes greater account of the social and environmental consequences of market behavior.

Decoupling Environmental Degradation and Economic Growth

Historically there has been a close correlation between economic growth and environmental degradation: as communities grow, so the environment declines. This trend is clearly demonstrated on graphs of human population numbers, economic growth, and environmental indicators. Unsustainable economic growth has been starkly compared to the malignant growth of a cancer because it eats away at the Earth's ecosystem services which are its life-support system. There is concern that, unless resource use is checked, modern global civilization will follow the path of ancient civilizations that collapsed through overexploitation of their resource base. While conventional economics is concerned largely with economic growth and the efficient allocation of resources, ecological economics has the explicit goal of sustainable scale (rather than continual growth), fair distribution and efficient allocation, in that order. The World Business Council for Sustainable Development states that "business cannot succeed in societies that fail".

In economic and environmental fields, the term decoupling is becoming increasingly used in the context of economic production and environmental quality. When used in this way, it refers to the ability of an economy to grow without incurring corresponding increases in environmental pressure. Ecological economics includes the study of societal metabolism, the throughput of resources that enter and exit the economic system in relation to environmental quality. An economy that is able to sustain GDP growth without having a negative impact on the environment is said to be decoupled. Exactly how, if, or to what extent this can be achieved is a subject of much debate. In 2011 the International Resource Panel, hosted by the United Nations Environment Programme (UNEP), warned that by 2050 the human race could be devouring 140 billion tons of minerals, ores, fossil fuels and biomass per year – three times its current rate of consumption – unless nations can make serious attempts at decoupling. The report noted that citizens of developed countries consume an average of 16 tons of those four key resources per capita per annum (ranging up to 40 or more tons per person in some developed countries). By comparison, the average person in India today consumes four tons per year. Sustainability studies analyse ways to reduce resource intensity (the amount of resource (e.g. water, energy, or materials) needed for the production, consumption and disposal of a unit of good or service) whether this be achieved from improved economic management, product design, or new technology.

There are conflicting views whether improvements in technological efficiency and innovation will enable a complete decoupling of economic growth from environmental degradation. On the one hand, it has been claimed repeatedly by efficiency experts that resource use intensity (i.e., energy and materials use per unit GDP) could in principle be reduced by at least four or five-fold, thereby allowing for continued economic growth without increasing resource depletion and associated pollution. On the other hand, an extensive historical analysis of technological efficiency improvements has conclusively shown that improvements in the efficiency of the use of energy and materials were almost always outpaced by economic growth, in large part because of the rebound effect (conservation) or Jevons Paradox resulting in a net increase in resource use and associated pollution. Furthermore, there are inherent thermodynamic (i.e., second law of thermodynamics) and practical limits to all efficiency improvements. For example, there are certain minimum unavoidable material requirements for growing food, and there are limits to making automobiles, houses, furniture, and other products lighter and thinner without the risk of losing their necessary functions. Since it is both theoretically and practically impossible to increase resource use efficiencies indefinitely, it is equally impossible to have continued and infinite economic growth without a concomitant increase in resource depletion and environmental pollution, i.e., economic growth and resource depletion can be decoupled to some degree over the short run but not the long run. Consequently, long-term sustainability requires the transition to a steady state economy in which total GDP remains more or less constant, as has been advocated for decades by Herman Daly and others in the ecological economics community.

A different proposed solution to partially decouple economic growth from environmental degradation is the *restore* approach. This approach views "restore" as a fourth component to the common reduce, reuse, recycle motto. Participants in such efforts are encouraged to voluntarily donate towards nature conservation a small fraction of the financial savings they experience through a more frugal use of resources. These financial savings would normally lead to rebound effects, but a theoretical analysis suggests that donating even a small fraction of the experienced savings can potentially more than eliminate rebound effects.

Nature as an Economic Externality

The economic importance of nature is indicated by the use of the expression ecosystem services to highlight the market relevance of an increasingly scarce natural world that can no longer be regarded as both unlimited and free. In general, as a commodity or service becomes more scarce the price increases and this acts as a restraint that encourages frugality, technical innovation and alternative products. However, this only applies when the product or service falls within the market system. As ecosystem services are generally treated as economic externalities they are unpriced and therefore overused and degraded, a situation sometimes referred to as the Tragedy of the Commons.

Deforestation of native rain forest in Rio de Janeiro
City for extraction of clay for civil engineering (2009 picture)

One approach to this dilemma has been the attempt to "internalize" these "externalities" by using market strategies like ecotaxes and incentives, tradeable permits for carbon, and the encouragement of payment for ecosystem services. Community currencies associated with Local Exchange Trading Systems (LETS), a gift economy and Time Banking have also been promoted as a way of supporting local economies and the environment. Green economics is another market-based attempt to address issues of equity and the environment. The global recession and a range of associated government policies are likely to bring the biggest annual fall in the world's carbon dioxide emissions in 40 years.

Economic Opportunity

Treating the environment as an externality may generate short-term profit at the expense of sustainability. Sustainable business practices, on the other hand, integrate ecological concerns with social and economic ones (i.e., the triple bottom line). Growth that depletes ecosystem services is sometimes termed "uneconomic growth" as it leads to a decline in quality of life. Minimizing such growth can provide opportunities for local businesses. For example, industrial waste can be treated as an "economic resource in the wrong place". The benefits of waste reduction include savings from disposal costs, fewer environmental penalties, and reduced liability insurance. This may lead to increased market share due to an improved public image. Energy efficiency can also increase profits by reducing costs.

The idea of sustainability as a business opportunity has led to the formation of organizations such as the Sustainability Consortium of the Society for Organizational Learning, the Sustainable Business Institute, and the World Council for Sustainable Development. The expansion of sustainable business opportunities can contribute to job creation through the introduction of green-collar workers. Research focusing on progressive corporate leaders who have integrated sustainability into commercial strategy has yielded a leadership competency model for sustainability, and led to emergence of the concept of "embedded sustainability" – defined by its authors Chris Laszlo and Nadya Zhexembayeva as "incorporation of environmental, health, and social value into the core business with no trade-off in price or quality – in other words, with no social or green premium." Laszlo and Zhexembayeva's research showed that embedded sustainability offers at least seven distinct opportunities for business value creation: a) better risk-management, b) increased efficiency through reduced waste and resource use, c) better product differentiation, d) new market entrances, e) enhanced brand and reputation, f) greater opportunity to influence industry standards, and g) greater opportunity for radical innovation. 2014 research further suggested that innovation driven by resource depletion can result in fundamental advantages for company products and services, as well as the company strategy as a whole, when right principles of innovation are applied.

Ecosocialist Approach

One school of thought, often labeled ecosocialism or ecological Marxism, asserts that the capitalist economic system is fundamentally incompatible with the ecological and social requirements of sustainability. This theory rests on the premises that:

1. Capitalism's sole economic purpose is "unlimited capital accumulation" in the hands of the capitalist class

2. The urge to accumulate (the profit motive) drives capitalists to continually reinvest and expand production, creating indefinite and unsustainable economic growth

3. "Capital tends to degrade the conditions of its own production" (the ecosystems and resources on which any economy depends)

Thus, according to this analysis:

1. Giving economic priority to the fulfillment of human needs while staying within ecological limits, as sustainable development demands, is in conflict with the structural workings of capitalism

2. A steady-state capitalist economy is impossible; further, a steady-state capitalist economy is socially undesirable due to the inevitable outcome of massive unemployment and underemployment

3. Capitalism will, unless overcome by revolution, run up against the physical limits of the biosphere and self-destruct

By this logic, market-based solutions to ecological crises (ecological economics, environmental economics, green economy) are rejected as technical tweaks that do not confront capitalism's structural failures. "Low-risk" technology/science-based solutions such as solar power, sustainable agriculture, and increases in energy efficiency are seen as necessary but insufficient. "High-risk" technological solutions such as nuclear power and climate engineering are entirely rejected. Attempts made by businesses to "greenwash" their practices are regarded as false advertising, and it is pointed out that implementation of renewable technology (such as Walmart's proposition to supply their electricity with solar power) has the effect opposite of reductions in resource consumption, viz. further economic growth. Sustainable business models and the triple bottom line are viewed as morally praiseworthy but ignorant to the tendency in capitalism for the distribution of wealth to become increasingly unequal and socially unstable/unsustainable. Ecosocialists claim that the general unwillingness of capitalists to tolerate—and capitalist governments to implement—constraints on maximum profit (such as ecotaxes or preservation and conservation measures) renders environmental reforms incapable of facilitating large-scale change: "History teaches us that although capitalism has at times responded to environmental movements . . . at a certain point, at which the system's underlying accumulation drive is affected, its resistance to environmental demands stiffens." They also note that, up until the event of total ecological collapse, destruction caused by natural disasters generally causes an increase in economic growth and accumulation; thus, capitalists have no foreseeable motivation to reduce the probability of disasters (i.e. convert to sustainable/ecological production).

Ecosocialists advocate for the revolutionary succession of capitalism by ecosocialism—an egalitarian economic/political/social structure designed to harmonize human society with non-human ecology and to fulfill human needs—as the only sufficient solution to the present-day ecological crisis, and hence the only path towards sustainability. Sustainability is viewed not as a domain exclusive to scientists, environmental activists, and business leaders but as a holistic project that must involve the whole of humanity redefining its place in Nature: "What every environmentalist needs to know . . . is that capitalism is not the solution but the problem, and that if humanity is going to survive this crisis, it will do so because it has exercised its capacity for human freedom, through social struggle, in order to create a whole new world—in coevolution with the planet."

Social Dimension

Sustainability issues are generally expressed in scientific and environmental terms, as well as in ethical terms of stewardship, but implementing change is a social challenge that entails, among other things, international and national law, urban planning and transport, local and individual lifestyles and ethical consumerism. "The relationship between human rights and human development, corporate power and environmental justice, global poverty and citizen action, suggest that responsible global citizenship is an inescapable element of what may at first glance seem to be simply matters of personal consumer and moral choice."

Peace, Security, Social Justice

Social disruptions like war, crime and corruption divert resources from areas of greatest human need, damage the capacity of societies to plan for the future, and generally threaten human well-being and the environment. Broad-based strategies for more sustainable social systems include: improved education and the political empowerment of women, especially in developing countries; greater regard for social justice, notably equity between rich and poor both within and between countries; and intergenerational equity. Depletion of natural resources including fresh water increases the likelihood of "resource wars". This aspect of sustainability has been referred to as environmental security and creates a clear need for global environmental agreements to manage resources such as aquifers and rivers which span political boundaries, and to protect shared global systems including oceans and the atmosphere.

Poverty

A major hurdle to achieve sustainability is the alleviation of poverty. It has been widely acknowledged that poverty is one source of environmental degradation. Such acknowledgment has been made by the Brundtland Commission report Our Common Future and the Millennium Development Goals. There is a growing realization in national governments and multilateral institutions that it is impossible to separate economic development issues from environment issues: according to the Brundtland report, "poverty is a major cause and effect of global environmental problems. It is therefore futile to attempt to deal with environmental problems without a broader perspective that encompasses the factors underlying world poverty and international inequality." Individuals living in poverty tend to rely heavily on their local ecosystem as a source for basic needs (such as nutrition and medicine) and general well-being. As population growth continues to increase, increasing pressure is being placed on the local ecosystem to provide these basic essentials. According to the UN Population Fund, high fertility and poverty have been strongly correlated, and the world's poorest countries also have the highest fertility and population growth rates. The word sustainability is also used widely by western country development agencies and international charities to focus their poverty alleviation efforts in ways that can be sustained by the local populace and its environment. For example, teaching water treatment to the poor by boiling their water with charcoal, would not generally be considered a sustainable strategy, whereas using PET solar water disinfection would be. Also, sustainable best practices can involve the recycling of materials, such as the use of recycled plastics for lumber where deforestation has devastated a country's timber base. Another example of sustainable practices in poverty alleviation is the use of exported recycled materials from developed to developing countries, such as Bridges to Prosperity's use of wire rope from shipping container gantry cranes to act as the structural wire rope for footbridges that cross rivers in poor rural areas in Asia and Africa.

Human Relationship to Nature

According to Murray Bookchin, the idea that humans must dominate nature is common in hierarchical societies. Bookchin contends that capitalism and market relationships, if unchecked, have the capacity to reduce the planet to a mere resource to be exploited. Nature is thus treated as a commodity: "The plundering of the human spirit by the market place is paralleled by the plundering of the earth by capital." Social ecology, founded by Bookchin, is based on the conviction that nearly all of humanity's present ecological problems originate in, indeed are mere symptoms of, dysfunctional social arrangements. Whereas most authors proceed as if our ecological problems can be fixed by implementing recommendations which stem from physical, biological, economic etc., studies, Bookchin's claim is that these problems can only be resolved by understanding the underlying social processes and intervening in those processes by applying the concepts and methods of the social sciences.

A pure capitalist approach has also been criticized in Stern Review on the Economics of Climate Change to mitigation the effects of global warming in this excerpt ...

"the greatest example of market failure we have ever seen."

Deep ecology is a movement founded by Arne Naess that establishes principles for the well-being of all life on Earth and the richness and diversity of life forms. The movement advocates, among other things, a substantial decrease in human population and consumption along with the reduction of human interference with the nonhuman world. To achieve this, deep ecologists advocate policies for basic economic, technological, and ideological structures that will improve the *quality of life* rather than the *standard of living*. Those who subscribe to these principles are obliged to make the necessary change happen. The concept of a billion-year Sustainocene has been developed to initiate policy consideration of an earth where human structures power and fuel the needs of that species (for example through artificial photosynthesis) allowing Rights of Nature.

Human Settlements

Sustainability Principles

1. Reduce dependence upon fossil fuels, underground metals, and minerals
2. Reduce dependence upon synthetic chemicals and other unnatural substances
3. Reduce encroachment upon nature
4. Meet human needs fairly & efficiently

One approach to sustainable living, exemplified by small-scale urban transition towns and rural ecovillages, seeks to create self-reliant communities based on principles of simple living, which maximize self-sufficiency particularly in food production. These principles, on a broader scale, underpin the concept of a bioregional economy. These approaches often utilize commons based knowledge sharing of open source appropriate technology.

Other approaches, loosely based around New Urbanism, are successfully reducing environmental impacts by altering the built environment to create and preserve sustainable cities which support sustainable transport. Residents in compact urban neighborhoods drive fewer miles, and have significantly lower environmental impacts across a range of measures, compared with those living in sprawling suburbs. In sustainable architecture the recent movement of New Classical Architecture promotes a sustainable approach towards construction, that appreciates and develops smart growth, architectural tradition and classical design. This in contrast to modernist and globally uniform architecture, as well as opposing solitary housing estates and suburban sprawl. Both trends started in the 1980s. The concept of Circular flow land use management has also been introduced in Europe to promote sustainable land use patterns that strive for compact cities and a reduction of greenfield land take by urban sprawl.

Large scale social movements can influence both community choices and the built environment. Eco-municipalities may be one such movement. Eco-municipalities take a systems approach, based on sustainability principles. The eco-municipality movement is participatory, involving community members in a bottom-up approach. In Sweden, more than 70 cities and towns—25 per cent of all municipalities in the country—have adopted a common set of "Sustainability Principles" and implemented these systematically throughout their municipal operations. There are now twelve eco-municipalities in the United States and the American Planning Association has adopted sustainability objectives based on the same principles.

There is a wealth of advice available to individuals wishing to reduce their personal and social impact on the environment through small, inexpensive and easily achievable steps. But the transition required to reduce global human consumption to within sustainable limits involves much larger changes, at all levels and contexts of society. The United Nations has recognised the central role of education, and have declared a decade of education for sustainable development, 2005–2014, which aims to "challenge us all to adopt new behaviours and practices to secure our future". The Worldwide Fund for Nature proposes a strategy for sustainability that goes beyond education to tackle underlying individualistic and materialistic societal values head-on and strengthen people's connections with the natural world.

Human and Labor Rights

Application of social sustainability requires stakeholders to look at human and labor rights, prevention of human trafficking, and other human rights risks. These issues should be considered in production and procurement of various worldwide commodities. The international community has identified many industries whose practices have been known to violate social sustainability, and many of these industries have organizations in place that aid in verifying the social sustainability of products and services. The Equator Principles (financial industry), Fair Wear Foundation (garments), and Electronics Industry Citizenship Coalition are examples of such organizations and initiatives. Resources are also available for verifying the life-cycle of products and the producer or vendor level, such as Green Seal for cleaning products, NSF-140 for carpet production, and even labeling of Organic food in the United States.

References

- White, F; Stallones, L; Last, JM. (2013). Global Public Health: Ecological Foundations. Oxford University Press. ISBN 978-0-19-975190-7.

- Barbier, Edward B. (2006). Natural Resources and Economic Development. //books. google.com/books?id=fYrEDA-VnyUC&pg=PA45: Cambridge University Press. pp. 44–45. ISBN 9780521706513. Retrieved April 8, 2014.

- Meadows, D.H., D.L. Meadows, J. Randers, and W.W. Behrens III. 1972. The Limits to Growth. Universe Books, New York, NY. ISBN 0-87663-165-0.

- Hazeltine, B.; Bull, C. (1999). Appropriate Technology: Tools, Choices, and Implications. New York: Academic Press. pp. 3, 270. ISBN 0-12-335190-1.

- Nussbaum, Martha (2011). Creating Capabilities: The Human Development Approach. Cambridge, Massachusetts and London, England: The Belknap Press of Harvard University Press. p. 16. ISBN 978-0-674-05054-9.

- Georgescu-Roegen, Nicholas (1971). The Entropy Law and the Economic Process. (PDF contains only the introductory chapter of the book). Cambridge, Massachusetts: Harvard University Press. ISBN 0674257804.

- Rifkin, Jeremy (1980). Entropy: A New World View. (PDF contains only the title and contents pages of the book). New York: The Viking Press. ISBN 0670297178.

- Lynn R. Kahle, Eda Gurel-Atay, Eds (2014). Communicating Sustainability for the Green Economy. New York: M.E. Sharpe. ISBN 978-0-7656-3680-5.

- Wandemberg, JC (August 2015). Sustainable by Design. Amazon. p. 122. ISBN 1516901789. Retrieved 16 February 2016.

- Hilgenkamp, K. (2005). Environmental Health: Ecological Perspectives. London: Jones & Bartlett. ISBN 978-0-7637-2377-4.

- Meadows, D.H., D.L. Meadows, J. Randers, and W. Behrens III. (1972). The Limits to Growth. New York: Universe Books. ISBN 0-87663-165-0.

- Botkin, D.B. (1990). Discordant Harmonies, a New Ecology for the 21st century. New York: Oxford University Press. ISBN 978-0-19-507469-7.

- Dalal-Clayton, Barry and Sadler, Barry 2009. Sustainability Appraisal. A Sourcebook and Reference Guide to International Experience. London: Earthscan. ISBN 978-1-84407-357-3.

- Bell, Simon and Morse, Stephen 2008. Sustainability Indicators. Measuring the Immeasurable? 2nd edn. London: Earthscan. ISBN 978-1-84407-299-6.

- Lutz W., Sanderson W.C., & Scherbov S. (2004). The End of World Population Growth in the 21st Century London: Earthscan. ISBN 1-84407-089-1.

- Groombridge, B. & Jenkins, M.D. (2002). World Atlas of Biodiversity. Berkeley: University of California Press. ISBN 978-0-520-23668-4.

- UNEP (2011). Decoupling Natural Resource Use and Environmental Impacts from Economic Growth. ISBN 978-92-807-3167-5. Retrieved on: 2011-11-30.

Permissions

Index

A

Acyl Carrier Protein (acp), 184

Animal Feces, 147-148

Animal Manure, 148-151, 184

Aquatic Environments, 232

Artificial Photosynthesis, 13, 32, 76, 89, 282

B

Bacterial Degradation, 142

Bio-energy with Carbon Capture and Storage (bec-cs), 233

Biochemical Oxygen Demand (bod), 169

Biodiesel, 2-3, 10, 16-17, 19, 106-107, 110, 179-180, 182-186, 191, 203-232

Bioethers, 186

Biomass Sustainability, 200

Biomass to Liquids (btls), 3, 16

C

Carbon Capture, 116-117, 233-237

Carbon Capture and Storage (ccs), 234, 237

Carbon Sequestration, 117, 122, 236

Caterpillar Gas Engines, 168

Chronic Kidney Disease (ckd), 159

Clean Development Mechanism (cdm), 236

Commercial Planting, 108

Concentrated

Photovoltaics (cpv), 31

Concentrated Solar Power (csp), 13-14, 20, 84

Concentrator Photovoltaics (cpv), 13

D

Depolymerization, 119

E

Economic Development, 70, 239, 242, 245-246, 255, 257, 259, 261, 281, 284

Efficient Fuels, 167, 169, 171, 173, 175, 177, 179, 181, 183, 185, 187, 189, 191, 193, 195, 197, 199, 201, 203

Electrogenic Micro-organisms, 5

Enhanced Geothermal Systems (egs), 30, 98

Environmental Protection, 145, 203, 213, 223-224, 239, 242, 252, 256-257, 259, 261

F

Food Crops, 107, 116, 221, 224, 226

Fossil Energy Ratios (fer), 182

Fuel Viscosity, 228

G

Geothermal Energy, 6, 14-15, 30, 92-94, 97-98, 100-101, 103

Global Warming, 8, 17, 25, 28, 53, 77, 93, 101-102, 111, 119, 169, 171, 188-189, 200, 268, 282

Green Building, 253, 257

Green Manure, 147, 149, 151

Greenhouse Gas (ghg), 7, 194

H

Heartwood, 128-131

High-density Polyethylene (hdpe), 214

Hot Dry Rock (hdr), 30

Human Manure, 149

Hydraulic Power-pipe Networks, 72

Hydroelectricity, 6, 13, 20, 46-47, 70, 72-74, 90, 271

Hydrogen-cell Power, 227

Hydropower, 6, 8, 12-13, 18-19, 29, 43, 62, 70-72, 74-76

J

Joint Implementation (ji), 236

L

Laminated Veneer Lumber (lvl), 140

Landfill Gas (lfg), 168

Leadership in Energy and Environmental Design (leed), 253

Levelised Cost of Electricity (lcoe), 20

Liquefied Natural Gas (lng), 115

Livestock Antibiotics, 151

Lpg (liquefied Petroleum Gas), 176

M
Microbial Gastrointestinal Flora, 194

N
Natural Biotic Systems, 240, 260
Nitrogen Fixation, 158
Non-methane Organic Compounds (nmocs), 173

O
Open-source Appropriate Technology (osat), 248
Organic Matter, 105, 119, 147, 149, 167, 169, 192, 195

P
Particulate Matter (pm), 17
Pellet Fuel, 167, 195
Permaculture, 245, 257, 273
Photosynthesis, 13, 32-33, 76, 78, 84, 89, 166, 179, 221-222, 230, 236, 270, 282
Plant Biomass Yields, 107
Polyethylene Terephthalate (pet), 84
Production Tax Credit (ptc), 51

R
Renewable Energy, 1-2, 6-11, 13, 15, 17-21, 23-29, 31, 33-35, 37, 41, 43, 45, 47, 49-57, 59, 62-63, 65, 67, 69-71, 73, 75-77, 79, 81, 83, 85, 87, 89, 91, 93, 95, 97, 99, 101, 103-104, 112, 116, 122-123, 125, 144, 163, 167-168, 174-176, 192, 201, 203, 220, 222, 234, 236, 244, 246-248, 257, 261, 271
Renewable Natural Gas, 105, 114-116
Reverse Electrodialysis (red), 31
Ribbon Cane Syrup, 160-161

S
Sapwood, 128-130
Sewage Biomass, 3
Social Development, 239, 242, 257, 259, 261
Solar Energy, 6, 10-11, 13-14, 46, 63, 76-81, 83-84, 86-87, 89-92, 98, 100, 103, 107, 163, 222, 245, 247, 260
Solar Water Disinfection (sodis), 84
Solid Biomass, 2
Structural Composite Lumber (scl), 140
Sub-tropical Convergence Zones (stcz), 236
Substitute Natural Gas (sng), 115
Sugarcane, 1-4, 10, 15-17, 24, 105-106, 110, 119, 124, 151-165, 179, 192, 235
Sugarcane Bagasse, 3-4, 164
Sustainability, 14, 26, 76-77, 91, 100-101, 200, 237, 239-240, 242-246, 248-253, 256-264, 266-267, 269-270, 275-284
Sustainable Agriculture, 245, 257, 259, 265, 269, 273, 279
Sustainable Development, 91, 239-246, 248-253, 255-259, 261, 265-267, 272, 276, 278-279, 283
Sustainable Development Goals (sdgs), 242, 265-266

T
Temperature Gelling, 215
Tidal Energy, 62, 65-66, 68-69
Tidal Turbines, 68
Transesterification, 10, 17, 110, 180, 182, 210-211, 216-217, 228
Transmission Network, 35

U
Ultra-low Sulfur Diesel (ulsd), 212
Urban Heat Islands (uhi), 86

W
Waste Vegetable Oil (wvo), 218
Whole-tree Harvesting, 5, 112
Wind Power, 6-7, 9, 12, 17-21, 29, 31, 33-38, 40-58, 60, 62, 247
Wind Turbines, 11-12, 31, 34-35, 37, 40, 42, 46, 48-55, 57-60, 64-65, 239, 271

www.ingramcontent.com/pod-product-compliance
Lightning Source LLC
Chambersburg PA
CBHW061932190326
41458CB00009B/2720